AMERICAN AVIATION ICONS 3

AF882401

CONVAIR

B-36 PEACEMAKER

WITH XC-99 & YB-60 DETAILS

THEODORE GIANNA

THUNDERBOLT 47 PUBLICATIONS

Front cover photo by USAF via National Archives

Copyright © 2022 Theodore Gianna. (THUNDERBOLT 47 PUBLICATIONS)
This publication is produced under the jurisdiction of international copyright laws. All rights reserved. No part of this publication may be reproduced, distributed, or transmitted in any form or by any means, including photocopying, recording, electronic, mechanical methods, or otherwise, without the express permission of the publishers and copyright holders.
Every effort has been made to make this book's information as complete and accurate as possible. However, the Publisher and Author accept no liability for any errors nor be held responsible for any subsequent use made of the published information, data, or specific details within this publication.

ISBN: 978-0-6456548-1-3 (Paperback)
First printing edition 2022.

INTRODUCTION

This is the third book in a series of iconic American military aircraft. It is the story of one of the most prominent bombers in US Air Force history; the Consolidated Vultee, later Convair, B-36.

The B-36 was the first bomber capable of delivering any nuclear weapon in the United States arsenal from within its four bomb bays without requiring any modifications. With a range of 10,000 miles and a maximum payload of 86,000 lb., the B-36 could fly to numerous global hotspots, including the Korean War, without refueling. However, the B-36 never did fight in any of the world's hot spots. Furthermore, it did not participate in the Korean War. Why? One reason the B-36 was not employed in Korea was that Korea was just a backwater, a strategic diversion from the actual menace on the horizon; the USSR, well, that was the official version.

For the Korean War, the Air Force relied on its dependable Boeing B-29s for bombing duties. Because of midair refueling, the Superfortresses could stay in the air for an eternity. Yet, incredibly, the B-36s were never equipped for in-flight refueling. Moreover, earlier B-36 variants were not maneuverable due to their excessive weight. The low maneuverability and the fact that their newly installed jet engines added further weight to the airframe and consumed so much fuel, the B-36's combat radius dropped to just over 3,000 miles.

Then, in 1954, Convair found its answer to the weight issues by making the B-36 a little faster and more maneuverable. The answer was simple: strip the aircraft of any non-essentials, and it would increase the bomber's speed and range. The project was named 'Featherweight.' Out went the extra gun turrets, half the crew, and other less essential items in the aircraft. As a result, the B-36s finally got some wind under their wings.

However, Convair's efforts were too little, too late. The Boeing marvel, the YB-52, was chosen to replace the B-36, while Convair's eight-engine jet bomber, the YB-60, fought against the YB-52 to become the next principal bomber in the USAF and lost that battle. These two events shook the organization. In light of this and the lucrative contracts for brand-new fighter jets, the company resolved to focus entirely on jet fighter production. As a result, air Force interceptor contracts for the F-102 and F-106 delta jets were awarded to Convair. However, despite its earlier promise to solely produce fighters, the company decided to manufacture the world's first supersonic jet bomber, the B-58 Hustler. Ultimately, the YB-60's loss to the YB-52 was the straw that broke the camel's back. Convair would not compete for another major USAF project after the B-58. Let's not forget that in the early 1960s, Convair's commercial airliners, the CV-880, and CV-990, also failed miserably.

ACKNOWLEDGMENTS

I want to thank Tony Landis, the AFMC History & Heritage Program, and Carswell Air Force Base for their photographs. I also want to credit the USAF, the National USAF museum, and the US Navy for their photos. Finally, I also want to thank Joe Baugher for his outstanding military website: http://www.joebaugher.com/. Highly recommended.

CONVAIR B-36 'PEACEMAKER'

DOUGLAS XB-19

Many believe the story of the B-36 began in 1941 when the Army Air Force (AAF) put in a request for a bomber capable of transporting 10,000 lb. of bombs over a distance of 10,000 miles. The proposal by the AAF was prompted by the worry that Axis forces would obtain triumph in Europe and Asia in the future. The AAF's 1941 request for a long-range bomber was like experiencing déjà vu.

We should go a few years back, before the war had even started, in 1935. During the fall of that year, the Air Corps submitted a list of specifications to several aircraft manufacturers for a bomber that would outperform all others in range, weight, and carrying capacity, similar to the B-36. In April 1936, Douglas was declared the competition's winner. The bomber was first designated as XBLR-2 (XBLR stood for 'Experimental Bomber Long Range'). Only one example would be built. The airplane was not any run-of-the-mill aircraft; it was the biggest airplane ever built in the pre-World War II era, the Douglas XB-19.

Contracts were signed in 1937, and Douglas engineers got to work developing the airplane and its components. However, from the start of production, the XB-19 had many problems. For instance, there were issues with the aircraft's electrical system, which only had the capacity of a small department store.

The XB-19 had a large monocoque wing (meaning the plane's skin is most of the structure, and the airframe supports the skin shape), the kind that had never been constructed before. The engineers were working without a precedent when they were developing the XB-19. They did not know how sturdy a massive structure like this would turn out. The designers later discovered that the larger the wing, the greater its carrying capacity per square foot of area. Had this been previously realized, a wing of a smaller area could have been used for the XB-19 proposed load. That is because aircraft with smaller wings could carry the same heavy load as larger winged planes if they traveled faster.

Because the enormous size and weight of the rudder, ailerons, and elevators of the XB-19 rendered direct operation through control cables impracticable, servo-tabs and hydraulic 'slave units' were attached to the surfaces, and the control was moved through the first five degrees of deflection via the servo-tabs. If the pilot needed more than five degrees of control movement, the hydraulic system kicked in, and mechanical 'slave units' responded to the pilot's commands. This was another first in hydraulic control boost technology. Although it was later updated, the hydraulic control was still employed on a variety of big aircraft as well as fast jet fighters long after the XB-19 left the scene.

The lessons in construction and aerodynamic design provided by the XB-19, nicknamed 'flying laboratory,' were carried forward into the B-29, B-36, and B-50. If not for the knowledge obtained during the manufacturing and operation of the XB-19, the legendary Boeing B-29 would have been manufactured far too late for use in the war against Japan. As for the B-36, it would also have taken more years of development before it would be ready.

The XB-19 was not just large; it was enormous. It measured 132 ft. in length, had a wingspan of 212 ft., and weighed 162,000 lb. on takeoff. The colossal aircraft was twice the size of the Boeing B-17 Flying Fortress and 30% larger than the Boeing B-29 Super Fortress.

The grand old bomber, nicknamed the 'Guardian of the Hemisphere,' was the world's largest plane until the construction of the Convair B-36 in 1946. Thus the Douglas XB-19 could rightly be referred to as the 'daddy of the super bombers.' Furthermore, it was a flying test stand for several engines, propellers, and accessory systems. The XB-19 was also the first aircraft to employ power-operated turrets for flexible weapons.

Regrettably, the XB-19 met its end in a very inglorious way. In 1946 it was laid up at Arizona's Davis-Monthan Airfield. Then, in 1950, for financial considerations, the USAF thought it was significantly less expensive to disassemble the XB-19 than it would have been to repair or simply maintain the enormous aircraft while it was in storage. Wright-Patterson Field's Air Material Command (AMC) Headquarters gave the order to destroy the enigmatic giant. The Air Force said that because it was non-airworthy, it could not be shipped due to its size or flown to the National Museum. AMC also said that technical development aircraft were not regarded as rare or historically significant types. Salvage was the final option. The XB-19's mission was over after nine years of valuable service.

The XB-19 'Flying Laboratory' in flight. (USAF)

A NEW AIR FORCE

One of the greatest Air Force generals in US military history, the unsurpassed General Henry H. Arnold, was a pioneer airman taught to fly by the Wright Brothers. He later became the commanding general of the Army Air Forces. He oversaw the nation's air operations during the global struggle against Germany and Japan both before and throughout World War II. Under his leadership, the air arm increased from 22,000 officers and men with 3,900 planes to approximately 2,500,000 men with 75,000 aircraft.

Shortly before his death in 1950, General Arnold expressed his belief that the best way to safeguard peace was to persuade potentially belligerent nations that the slightest aggressive action would instantly set in motion an intercontinental strategic force capable of causing the total destruction of the belligerent state.

General Mitchell, a sage and pioneer of the USAF in the early 1950s, was on the same wavelength as General Arnold. He stated that the US should create an Air Force capable of the greatest possible range, taking into account the other requirements, such as the capacity needed, the necessary defensive armament, and the efficiency and capability of the crews. It was obvious that the ideas of General Mitchell were crystal clear, and the goals he espoused were reached with the introduction of the Convair B-36.

THE SEARCH FOR A BOMBER

The supreme leader, General Henry H. Arnold. (USAF)

It all began in 1941. Pessimistic forecasts of the British Empire's demise had taken root in the minds of American strategists; safeguarding the Western Hemisphere from fascism would thus become the duty of the previously uncommitted United States. One possible answer was to build a long-range bomber capable of flying across the Atlantic Ocean to central Europe to bomb critical Nazi infrastructures.

Thus, in 1941, at the United States Army Air Corps (USAAC) request, a crucial meeting was organized with the participation of decision-makers, strategists, and manufacturers. There were a lot of general, tactical, and technical debates during the session. Convair, Northrop, Douglas, and Boeing were all present. All the manufacturers received the relevant specifications

Initially, the USAAC required a bomber capable of carrying 4,000 pounds of bombs over 12,000 miles. However, in August 1941, a second conference held in Washington resulted in the final program of 10,000 pounds of bombs at 10,000 miles, leading to the catchy slogan of the 'ten-ten bomber.'

XB-36

In November 1941, Convair was announced as the winner of the contract. The company was tasked to build two XB-36 prototypes for $75 million each. It marked the start of the most intensive research and development work on a military aircraft ever undertaken. Immediate testing of materials and machining procedures was conducted. During manufacturing and assembly testing, magnesium, polymers, and aluminum were utilized specifically for the B-36.

Several years of arduous work would be required to produce the B-36. The technology in the early 1940s was not ready to build such a gigantic aircraft as the B-36. Therefore, it was necessary to wait for technological advancement in order to benefit from it without slowing down the aircraft's progress. In 1942, Convair's engineers became obsessed with the formidable distance to be covered and, consequently, with the aircraft's weight. The ratio of the structure's weight to weight carried became the primary design requirement. The theoretical maximum range for an airplane in 1942, regardless of size, was presumed to be 10,000 miles. In reality, it was not so simple. However, in November 1938, three British Vickers Wellesley aircraft set a distance record when they flew from Ismailia, Egypt, to Darwin, Australia, a distance of 7,162 miles. They succeeded only because all three planes carried gasoline on board, nothing more. Convair would need to build a bomber that could fly 10,000 miles with a 5-ton bomb load, a large amount of heavy defensive weaponry, and a crew of fifteen. That would have been spectacular in 1942. It was estimated that adding ten pounds to the aircraft's weight would add one mile to the total distance; adding one ton would add 200 miles! Furthermore, three additional requirements would be needed to successfully get the B-36 to its destination: minimal drag, minimal fuel consumption, and maximum engine endurance. Convair's attempt to accomplish the feat at the time would have been extremely difficult, if not impossible.

Aerodynamic problems captivated Convair's technicians. The double fins were abandoned in favor of a giant axial fin. Wingtip stall slots were fitted to the first two wing panels of the XB-36s after the National Advisory Committee for Aeronautics (NACA) tunnel tests exhibited poor stall characteristics. But the cracks were later sealed when it was proven in flight that the B-36 stalled very well despite the NACA tests. Two-speed cooling fans were made to keep the temperature of the fully enclosed motors within allowable limits. A two-speed propeller drive relay was considered but soon abandoned when it became apparent that the system would require several years of study. In this case, it was a question of doubling the action of the variable-pitch propellers by changing the speed on the shaft. Propeller vibrations, a fundamental difficulty of rear thrusters, were studied with particular care. A complete set of engine nacelles, placed on the corresponding part of the wing structure, was tested in the wind tunnel. At the same time, the gutters for bringing air to the engine from the wing's leading edge were developed. Before the B-36 was released, it was evident that modifications would be needed as the production models entered the 1,600-meter-long final assembly line. As a result, the second XB-36 was changed to a YB-36 service test aircraft identical to the production models. Its nose section was modified, and instead of a flat windshield continuing the general line of the back of the fuselage, a protruding 'blister' window was fitted so that the flight engineer could look back and watch the propellers without having to change places.

When Convair's engineers were working out the aerodynamic issues, the war in Europe was suddenly less threatening for the allies as it had swung their way. Thus, the need for the B-36 was diminishing. Nevertheless, the United States Army Air Forces (USAAF) declined to condemn the B-36 conclusively since the US was fully involved in the war against Japan and could not rule out the possibility of defeat in that theater of warfare. Moreover, the B-36 could be used to bomb Japan from Hawaii. Its development was, however, hampered by the difficulties inherent in a war economy. Because manpower and raw materials were scarce, higher priority was given to other aircraft that were promptly called into war. Nonetheless, in late 1943, The B-36 received final approval from the AAF, and Convair was awarded a contract for 100 aircraft.

The B-36 was the largest aircraft ever ordered in America and the first mass-produced propeller-driven aircraft. In 1942, the main sections of the first XB-36 left San Diego and were transferred by car flatbed to the government plant in Fort Worth, Texas, for assembly. A freighter variant of the bomber, the XC-99, was also approved for construction in San Diego in 1942.

The XB-36, 42-13570, made its inaugural flight on August 8, 1946. It had massive 110 in. wheels, which required specially prepared runways. The two main wheels were changed to smaller, four 56-in. wheels in consequent variants. The second prototype, YB-36, 42-13571, which had a bubble canopy, made its first flight on December 4, 1947, and was converted to a YB-36A on May 27, 1948.

The single prototype, Convair XB-36, Model 36, 42-13570, in flight. It was powered by six 3,000 hp, Pratt & Whitney R-4360-25 28-cylinder four-row radial piston aircraft engines. The bomber was unarmed, but provisions were made for five 37 mm cannon, ten 0.5in. guns , and up to 72,000 lb, of bombs. This was the version without the bubble canopy. (T. Gianna, USAF)

The XB-36 prototype coming in for landing with its giant 110 inch wheels. (T. Gianna, USAF)

The prototype Convair XB-36 parked beside a Boeing B-29-55-BA,44-84027, for size comparison at Carswell Air Force Base, Ft. Worth, Texas in 1948. (USAF)

XB-36 flying over lush countryside. (USAF)

THE CONTROVERSIAL BOMBER

There were certain opponents of the B-36 who wanted the project canceled. The strategic super-bomber was on shaky ground and was the subject of numerous controversies that rattled the military and even the US government's highest political circles. General Nathan F. Twining, for instance, desired a more effective aircraft and urged halting production of the B-36As while waiting to develop more effective models. However, the B-36's most ardent opponent was the commander of the Strategic Air Command (SAC), General George C. Kenney, who urged General Spaatz to cancel the B-36 order altogether. Several others were more cautious but not outright antagonistic to the B-36 program as the leader of SAC.

However, the biggest challenge to the approval of the B-36 came from one of the other services, the US Navy. In 1947, the USAF separated from the Army and was put under the administration of the newly formed National Military Establishment.

In the spring of 1949, a major controversy around the B-36 and questions of service unification developed. Officially, the argument was about the B-36 acquisition process, but it expanded to encompass a discussion of the Air Force's and Navy's duties and missions. The Air Force generals of the newly created Air Force put forth a new doctrine, 'strategic bombing,' by using nuclear weapons. The generals claimed that was the only critical element needed to win any future war and, therefore, the only way to prevent an enemy from launching a surprise attack against the United States in the manner of Pearl Harbor. The Air Force generals claimed that substantial finance for this project, starting with an improved B-36 Peacemaker intercontinental bomber, was necessary.

The Navy's admirals had different opinions. They requested funding for a large fleet of 'Supercarriers' and their accompanying battle groups, starting with the USS United States, citing the overwhelming dominance of the aircraft carrier in the Pacific Theater in World War II. Strategic bombing alone, with or without the deployment of nuclear weapons, was not the Navy's idea of how to win wars. The Navy also argued that it was unethical to decide to use atomic weapons extensively at the commencement of any future battle by bombing the civilian population centers of the enemy country.

It looked like the Navy had the upper hand, as the Secretary of Defense, former Secretary of the Navy James Forrestal, supported the Navy's position and authorized the construction of the USS United States with a production run of five ships. Unfortunately, he was fired by President Truman on March 28, 1949, following a series of budgetary disagreements. He was replaced by Louis A. Johnson, a strong supporter of Truman's budget limits and an ardent supporter of the new nuclear bomber force of the Air Force.

Johnson consequently canceled the 'Supercarriers' project of the US Navy while firmly supporting the Air Force's B-36 bomber program. Unfortunately, this action led to a terrific feud between the services, in which Navy personnel leveled dark accusations of misconduct at Air Force officials. In the ensuing feud over nuclear weapons-equipped bombers, a chief of naval operations was driven out of office. Moreover, Secretary of the Navy John L. Sullivan and other high-ranking admirals resigned.

Op-23 was a research group led by Captain Arleigh A. Burke, the distinguished Admiral in World War II and the Korean War. He started gathering data critical of the B-36's performance and capabilities. As a result, an 'anonymous paper' quickly surfaced, saying that the B-36 was a 'billion-dollar disaster' and accusing B-36 contractors of fraud. According to the paper, Johnson, who served on the board of directors of Convair, the B-36's manufacturer, had a personal stake in its production.

After all, was said and done, the USAF won the day, thanks mainly to the support of Johnson. Of course, the US Navy had valid reasons for funding the 'Supercarriers' project and was ethically in the right on the use of nuclear weapons on heavily populated centers by the Air Force, but victory was not going to go the Navy's way when Johnson was calling the shots.

The reasons for the Navy's loss were because of poor leadership and an inefficient public relations campaign. On the other hand, the Air Force displayed strong leadership and effective PR. Although that didn't make the Air Force's proposed doctrine right, it was bought by the public because the Air Force was just more prepared and savvier in the eyes of the people. The failure of the Navy to successfully promote its carrier doctrine to policymakers and the Air Force's selling of an inflexible and unproven ideology caused a lot of hostility between the two services, which would negatively impact America's preparedness for future conflicts.

TECHNICAL DESCRIPTION OF THE B-36

There was one question in many people's minds. Why did the Convair B-36 have its propellers behind the wings, unlike almost every other propliner in the early 1950s? The answer was given by the National Advisory Committee for Aeronautics (NACA). First, the NACA did various tests in laminar-flow wing profiles (the smooth, uninterrupted flow of air over the contour of the wings) and on the B-36's engine, the Pratt & Whitney 28-cylinder, 3,000 hp. 'Wasp Major,' with positive results. Subsequently, the laminar-flow airfoil was mounted slightly forward at the fuselage's mid-point. Finally, the NACA decided to mount the engines as pushers because that arrangement simplified the wing structure by eliminating the need for nacelles and offered the benefits of reduced noise.

An airplane as large as the B-36 had never previously been attempted, including the XB-19. As a result, almost all the components needed for the plane had to be specially designed and manufactured. Items like the landing gear, the wing spars, the fuselage frames, the fuel systems, and the electrical systems were created from scratch. At the same time, the electrical and fuel systems, landing gear, wing spars, fuselage frames, and other components were outsourced to independent manufacturers to make the parts.

The primary goal of an airplane required to fly with a massive load over such incredible distances was to be lightweight. Consequently, magnesium was used for the first time in large aircraft construction. This novel, lightweight material was utilized in the fuselage side panels and the wing trailing edge. In addition, magnesium plates were bonded together in a waffle design instead of rivets in order to reduce weight further.

Approximately 1,500 lb. was saved by switching to an alternating current electrical system compared to a direct one. Furthermore, it made the anticipated electrical issues at 40,000 ft. easier to handle by allowing induction motors to eliminate slip rings (conducting rings) and high-altitude arcing.

Instead of the more traditional semi-monocoque structures, the bomber bay portion of the fuselage was built on a truss system (an assembly of beams connected by nodes that creates a rigid structure) which consisted of two hefty keel structures.

The cabin pressurization problem was reduced to a forward and an aft part. The pressurized forward and aft cabin space totaled 3,924 cubic ft. Extending the length of the entire bomb-bay section, on the left side of the fuselage below the wing, was a pressurized magnesium communications tunnel 85 ft. long and two ft. in diameter. This tunnel connected the forward and aft cabins and was provided with a 4-wheel scooter which the crew members used to travel from the front cabin to the rear cabin and vice-versa.

Two decks made up the B-36's nose section. The bombardier and navigator were housed on the lower deck in a glass-enclosed compartment in the nose, while the pilot, copilot, and flight engineer had stations on the top deck. The radio/radar communications operators were just behind in a roomy compartment. The remaining crew members could be found behind the wings, in six-bunk housing. A B-36 crew consisted of 15 men, of which 11 were always on active duty, and four were constantly on standby.

The propellers of the B-36 were the biggest ever constructed at the time, measuring 19 ft. in diameter. Curtiss Electric's three-blade, hollow-steel propellers were full-feathering and reversible, with an inbuilt anti-icing system. In addition, the wing's leading edges were double-skinned to channel 'anti-icing' hot air. The hot air was also used to thaw the pilot's and bombardier's enclosures and the gun-sighting blisters.

At their base, the wings measured 7.5 ft. in depth. A walkway enabled crew members to access the six engine nacelles for inspection and servicing while the aircraft was in flight. The thickness of the wings' skin ranged from 0.375 inches at the base to 0.020 inches at the tips. The wings held 21,116 gallons of gasoline and 1,200 gallons of oil in their six integral fuel compartments.

The armament system had five remotely-operated turrets armed with 20 mm cannon. The nose turret was equipped with two cannons, while the upper front, upper aft, and lower aft turrets each housed four cannons. The tail turret, however, was equipped with two cannons. In the tail turret, provisions were also made for a radar-controlled 20 mm cannon.

The tail turret with the two AN/APG-3 radar controlled 20 mm cannon. (USAF)

This is one of the two giant 110 inch wheels of the XB-36. (USAF)

The two dorsal turrets, each with two 20 mm cannon. (USAF)

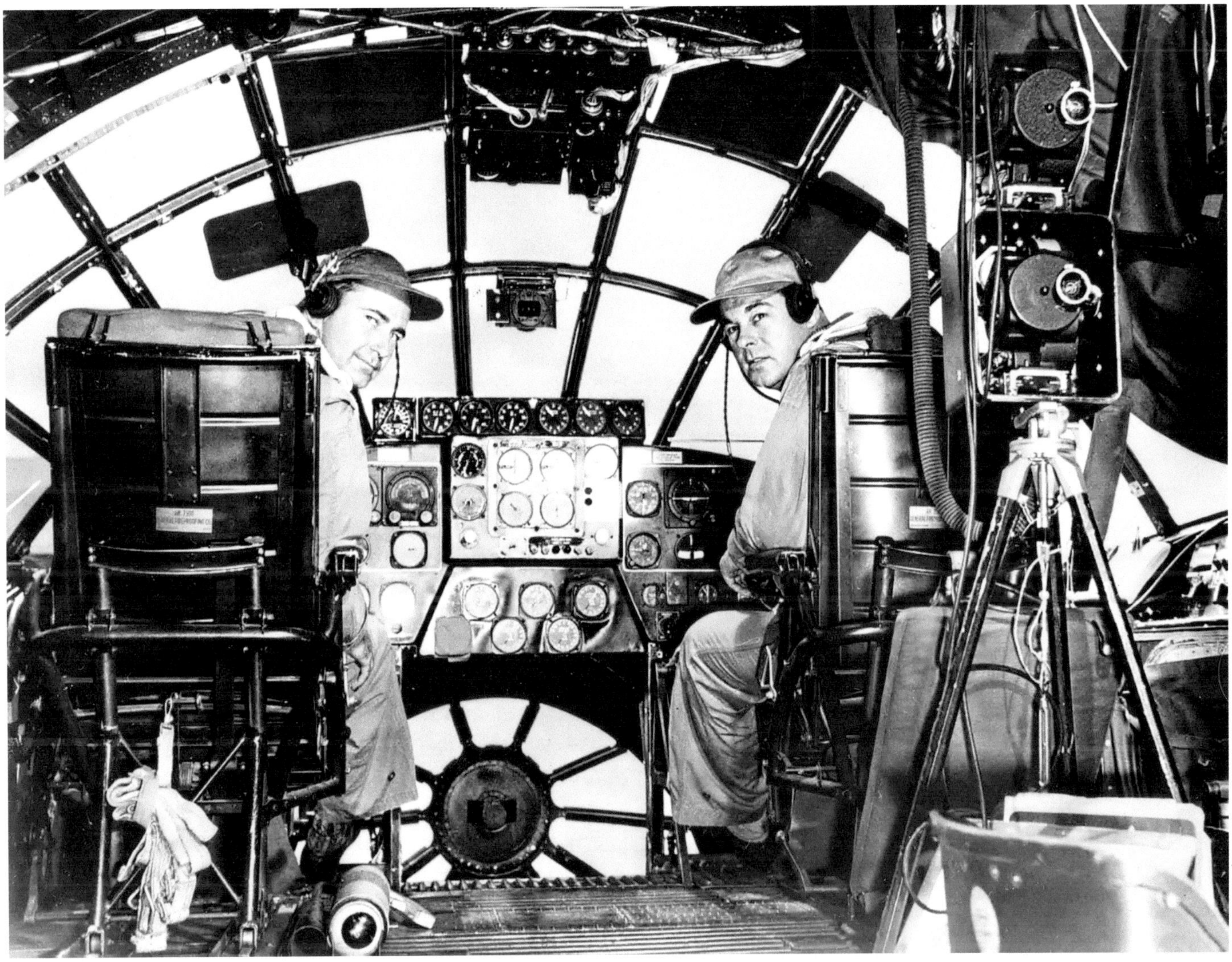

The two Convair test pilots at the controls of the XB-36 prototype. On the left is Beryl A. Erickson, and on the right is Gus Green. There were seven other crew members on the plane for its first flight on August 8, 1946. (USAF)

The B-36's propellers were 19 ft. in diameter, and were the largest propellers ever made, at the time. They were manufactured by Curtiss Electric, which was a part of Curtiss Wright . (USAF)

THE ENGINE: PRATT & WHITNEY R-4360 WASP MAJOR

The 3000 hp Pratt & Whitney R-4360-4 Wasp Major piston engine made its first run in 1944. Production began that same year and did not cease until 1955, when 18,697 R-4360s were produced. It was a four-row, 28-cylinder engine that wasn't used until close to the war's end. The R-4360 Wasp Major was fitted on late-model Boeing B-29s that were, in essence, early Boeing B-50s.

Cooling the rear cylinders of a four-row engine would have seemed formidable a few years earlier, but in the early 1940s, Pratt & Whitney made it appear easy. Each row of seven cylinders was staggered at a diagonal angle behind the one in front. This placement gave each bank of four the appearance of a canted inline. The back side of each bank was closely fitted; the front side was less so. This design worked very well. It allowed each of the seven magnetos to be placed in front of the bank it served. Despite the effect this must have had on cooling airflow.

The fascinating thing about this engine was that the air passing through the space between the front cylinders was adequate enough to cool the 21 cylinders in the remaining rows. Well, at least, that was the case in flight. However, the R-4360 tended to overheat when it was on the ground. It was one of only a few engines that could be shut down in perfect operating condition and have some problem with it the next time it was started up. The Franklin Automobile Company discovered in 1930 that the most effective cooling for the smallest amount of air was obtained if the air passed over the intake port (the final part of the engine's air intake system) before cooling the exhaust. Thanks to that breakthrough, the R-4360 was the first aircraft engine to incorporate this feature in its design.

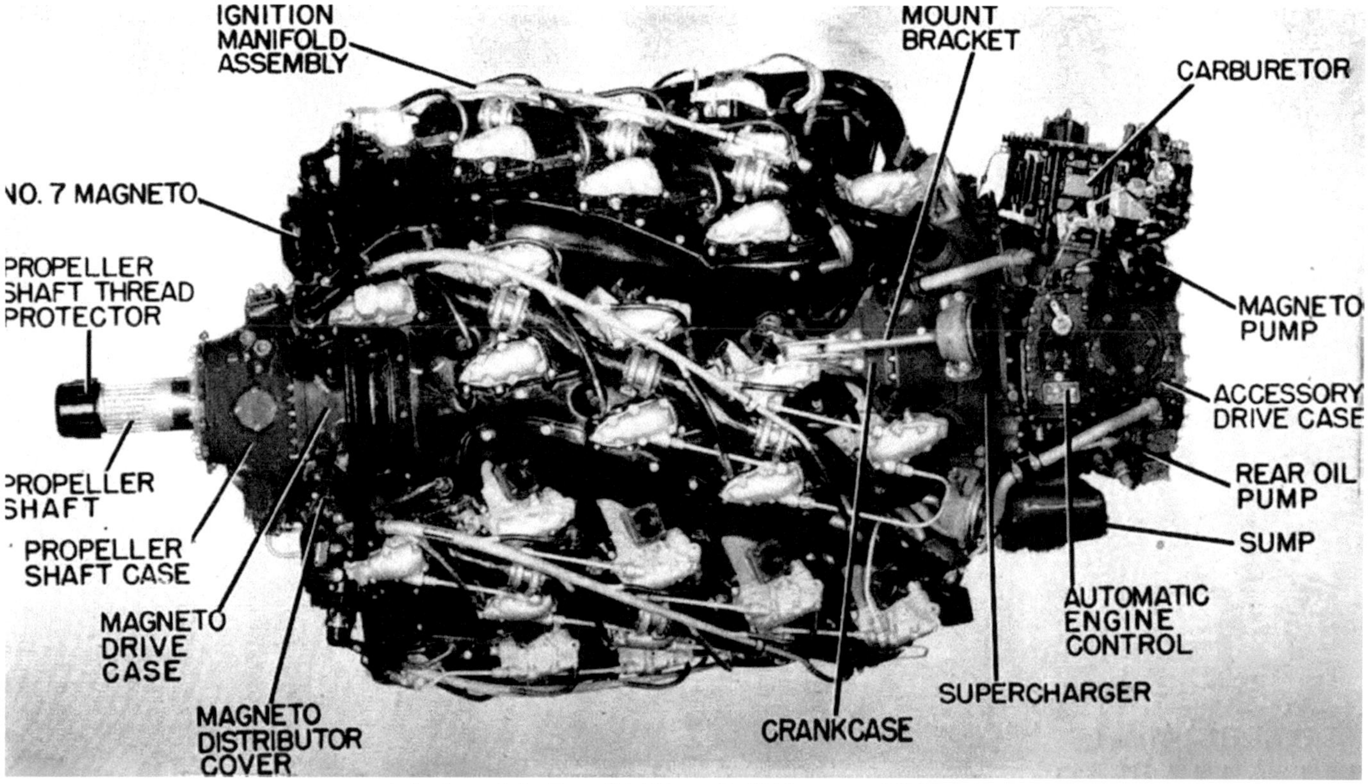

Left view of the Pratt & Whitney R-4360 engine. (US Navy)

The USAAF designed a sequence of tests so elaborate for the B-36 that the shutdown procedure took 6 hours, and the start-up operation took 4 hours. The R-4360 enabled the Bendix Engine Analyzer to demonstrate its use. This system used a cathode-ray tube (which produces images when an electron beam strikes a phosphorescent, or glowing, surface) to monitor nearly every aspect of a running engine. When tested on the B-36, it demonstrated a method for simplifying the grueling shutdown procedure. Unfortunately, the Bendix Engine Analyzer was a highly complicated device. It required a sensor and corresponding wire for every operating characteristic it monitored. For instance, each spark plug had to be connected to the analyzer, which required 56 wires per engine. Given the circumstances, it is hardly surprising that the engine analyzer did not become standard equipment on the B-36.

The R-4360 Wasp Major presented significant technical challenges. The four-row arrangement had significant temperature issues that reduced dependability, necessitating a rigorous maintenance regime that included regular cylinder replacement. Additionally, the necessity for big cooling flaps reduced aerodynamic efficiency, increasing the demand for engine power when cooling requirements were greatest. As a result, almost all aircraft equipped with the R-4360 were costly to operate and suffered decreased availability. However, one fortunate aircraft had fewer issues with the R-4360. The B-36 was one of the few shaft-driven aircraft to succeed with the Wasp Major. The large R-4360s were concealed in the thickest portion of the B-36'S wings and powered the pusher propellers through lengthy extension shafts extending rearward. The B-36's large number of cylinders was the likely reason for its fewer engine problems than other aircraft.

This is a Pratt & Whitney R-4360 Wasp Major at Oregon Air and Space Museum - Eugene, Oregon. Note the small B-36 model on the right side of the engine. (Photo by Daderot)

TRACK GEAR

The Air Force spent a lot of time experimenting with ways to set up forward airstrips for more efficient operations against an enemy. However, building and maintaining front-line strips was risky, especially when the runways needed to be lengthy and firm to support the enormous wheel loads of heavier aircraft.

The Fairchild Aircraft Division tested the feasibility of adding track gear to heavy cargo aircraft using a C-82 Packet airframe in the fall of 1945. The performance of the gear designed by Firestone exceeded the expectations of the gear's designers. The gear was composed of a belt, wheels, brakes, a shock-absorbing frame, and a retracting mechanism made of magnesium.

The 'track gear' had the advantage of functioning from sand, bare ground, and sod while sinking less than one-third as much as tires. Tundra, ice, and wind- blown snow areas would also be accessible. Furthermore, the heaviest planes flying from existing runways would have their weights dispersed over roughly 2.5 times the regular wheel-contact area.

It was too good to be true, and a USAF contract for 'track gear' was approved in November 1946 for $250,000. Even General George C. Kenney of the Strategic Air Command, who would later attempt to derail the B-36 project, was eager to test an entire task force equipped with track treads.

During the summer of 1948, things started to fall apart with the 'track gear.' First, the Goodrich belts for experimental gear testing turned out to be too weak, according to initial tests. Then, when the gear was tested in the sand, structural failure occurred in the nose gear track, and the main gear belt ripped off. Unfortunately, the production gear could not be stopped for the test results and had to keep on being produced on schedule.

In 1949, the Air Force Material Command (AMC) disclosed that it had fallen behind in the track gear testing phase. That was because it had to return the experimental equipment to Fairchild while production models were still being built. This step delayed the testing of the experimental set. The AMC stated that if the experimental equipment had been tested sooner, it would have canceled the project.

It was proven that gear maintenance under varying ground conditions was a significant issue. The track gear functioned well on sod and some mud, but it couldn't work on sand that didn't have a certain amount of dirt. It was also agreed that ski equipment was superior for snow operations. With further testing by AMC, the decision was made not to progress with the track gear system for the C-82. Boeing also used the track gear on the B-50. The results of the tests were mixed, and Boeing stated it would not require further testing on the track-equipped B-50. Despite the track gear issues on the C-82 and B-50, the track gear was tried again on the Air Force's largest aircraft at the time, the Consolidated B-36.

The Consolidated contract was modified in April 1948 to provide for the design, production, and installation of the B-36 'track gear' system. The first XB-36, 42-13570, was assigned to the test program. However, due to previously scheduled XB-36 testing, the track gear program was postponed until November 1949. Finally, in February 1950, the 'track gear' installation was completed, and the test program began with a weight restriction of 250,000 pounds.

After completing taxi testing in early March 1950, the 'Peacemaker' was prepared for its flight using the track gears. The XB-36's 'track gear' setup was the biggest system ever tested on an aircraft. Unfortunately, only one test was done, which took place on March 26, 1950, and that did not go well. The takeoff roll of the B-36 was regarded as rough and loud. Once in the air, the bomber circled the airport once before coming in for a landing. After touching down, parts of the track system left a trail of debris down the runway. The track system was judged totally unsuitable for such a large aircraft as the 'Big Stick.' It was no surprise that the project was eventually scrapped.

Initially, the 'track landing gear' concept was quite intriguing. Even though the tests were unusual, the USAF believed they were worthwhile. However, it was one of those concepts that looked excellent on paper but failed when put through rigorous testing. As a result, the USAF had to admit that the 'track gear' system was a complete failure and not worth continuing. So it was back to using wheels again for the Convair B-36.

The prototype XB-36 on the mat with its set of 'track gears'. The Peacemaker was far too big, and too heavy for the track gears. It was a good thing that the USAF abandoned the project. (T. Gianna, USAF)

This is the right main 'track gear' on the B-36. (USAF)

This is the front 'track gear' (USAF)

THE ACCIDENT THAT SAVED THE B-36 PROGRAM

On Friday, March 26, 1947, the prototype XB-36 had just received 11,000 gallons of fuel, causing its weight to increase to 253,998 lb. The test pilot. Beryl Erickson was the first to board with his offsider, test pilot Gus Green, right behind him. The crew included three technicians from Air Materiel Command (AMC): Lieutenant Wister G. Williams, Sergeant Major Dillon, and W. Mathews. There were also two Curtiss-Wright company employees: Lewis Clark and Arthur D. Nordheim, and two flight engineers, Jim Mac Eachern and William Easley.

At 12:20 pm, the XB-36 prototype was cleared for takeoff. The weather was mostly cloudy, with a temperature of 71.2°F, visibility at 14.9 miles, and a headwind of 21.9 mph. The big bomber took off at a speed of 130 mph.

The pilot raised the landing gear, and as it almost reached its fully retracted position, an extremely violent eruption shook the whole aircraft. The hydraulics blew away the huge weight of the right main gear with its massive 110-inch diameter tire and 800 lb. innertube. The retracting hydraulic strut ruptured, allowing tons of suddenly unsupported dead weight to plummet to the ground. As fuel and hydraulic lines broke, the back of the number four engine nacelle was destroyed. The drop also pulled the primary gear brace off the enormous wing spar.

Without the brace, the right main gear swung back and forth in the aircraft's slipstream like a massive rubber-tired swing. The bulk was prevented from swaying sideways by the oleo struts of the gear and a smaller, undamaged extension rod.

There were sounds of rending metal under tremendous stress, violently shrieking and rupturing. It was as if the plane had flown into hell. Erickson later stated that from the sounds and the way the plane was wobbling, he thought they had suffered a mid-air collision.

Through the intercom, one of the men at the back of the plane reported that smoke and even small, rapid flames were emerging from engine no. 4. Erickson decreased the power, turned off the engine, and feathered the propeller. Fortunately, the fire extinguishers responded instantly, and the fire was quickly extinguished.

As the aircraft climbed to 10,000 ft., the two pilots, Major Dillon and the two mechanics debated what to do next. They all agreed that an immediate landing was out of the question due to the plane's heavy load.

Erickson was confident that he and Green could land the damaged bomber, but with twelve additional passengers, that complicated matters. The greatest danger was that the landing gear would fail, allowing the right wing to crash to the ground and separate, burning the 21,000 gallons of very flammable aviation gas on board. Without a mechanism to discharge the gasoline, the pilots were forced to fly around and reduce the fuel level.

The pilots tried flying the enormous bomber using various rudder, aileron, and engine power combinations over the next six hours while radioing Convair about their intended landing procedure.

Major Dillon was instructed to bail out first, then return as quickly as possible to the airfield and find a vantage point from which he could radio steering instructions to Erickson after the landing. Dillon jumped but sprained his ankle as he landed. Eventually, he found a Douglas C-47 with a working radio.

Two crew members would jump and parachute down at each pass while Erickson flew around the drop zone at 5,000 ft. in a wide circle. Unfortunately, the winds were gusting between 30 and 35 miles per hour, and the bailout would be dangerous. The injuries sustained by nine of the twelve crew who parachuted ranged from sprained ankles to fractured ribs, shattered legs, and punctured lungs.

Meanwhile, the XB-36 continued to turn and burn up its fuel. Finally, at 3:45 pm, Erickson reported that he was at 6,000 ft. At 5:30 pm, he informed the control tower that the B-36 was much lighter and it would be possible to undertake the dangerous landing. However, before landing, he had to wait another 45 minutes for the strong wind to die down.

Due to burst hydraulics, Erickson and Green attempted to land an aircraft with no brakes and no nose wheel steering. To guide the massive plane, they would have to juggle rudder and aileron controls, as well as the power of the five remaining engines. At 6:15 pm, Erickson made a final turn over Lake Worth and began his approach with a 25 mph headwind.

With a straight landing gear extended and unlocked on a massive plane that he still knew little about; Erickson didn't have much choice. He made the decision to touch the plane down on the runway's grassy crash strip on its wheels. The significant challenge was to delicately adjust the ailerons in order to avoid damaging the left landing gear or putting too much pressure on the failing right landing gear. All of the weight was on Ericson's shoulders. He was well aware that if he crashed the plane, the crew's lives would be lost, and the B-36 project would end in failure. With nerves of steel, he brought the aircraft down the grassy strip and used stick and rudder to vary the weight on the right wheel and keep it in place. The aircraft rolled to the end of the runway and edged off onto the soft earth at the left, and stopped, but the right main gear held, and the wing never touched the ground. That was fortunate because there were still 8,000 gallons of high-octane fuel on board.

The two pilots saved not only their own lives but the future of the B-36 program itself. Its development was in peril because it was running late and over budget. Moreover, the Navy, which preferred more aircraft carriers than strategic bombers, was heavily criticizing the price of a fleet of strategic bombers. Convair hosted a testimonial luncheon in honor of Erickson and the crew as the company was relieved that the B-36 program was back on track. Convair President Harry Woodhead even presented all the crew members with engraved gold watches.

As for the XB-36, it flew again two months later after having new, redesigned gear retraction struts fitted. Not only that, but the XB-36 also broke a record by taking off at a gross weight of 278,000 lb., the highest load ever lifted into the air. The giant bomber stayed aloft for 5 hours for 'maximum gross' tests. Then on August 14, 1947, it created another record by reaching an altitude of 37,400 ft.

Regrettably, the Convair XB-36's final flight occurred on August 8, 1950. The military quickly concluded that the Wright Paterson AFB-based prototype, which was highly experimental and had undergone significant modifications since its inception, was no longer useful.

In October 1951, the XB-36 was transported back to Fort Worth. It was formally taken out of service on January 30, 1952. Its debris lingered near the northern side of the Fort Worth field for five years until being hauled to Carswell AFB in May 1957 and utilized for fire drills. It was a gloomy ending for the giant bomber.

B-36A/ B-36B

Series production in the gigantic Fort Worth factory started very early, so the first series was built simultaneously with the second prototype, the YB-36. Designated B-36A, the production model had the definition of the YB-36 and took up practically the same elements, except the landing gear, which consisted of bogies. The first production aircraft, B-36A, 44-92004, coded BM-004, was completed in July 1947, and on August 28, 1947, Beryl Erickson took it to the air for its first flight.

The first B-36A was virtually identical in design to the earlier modified YB-36A. For a while, the first B-36A and the following 12 aircraft were referred to as YB-36As. Except for the first B-36A, 44-9004, which was condemned for static testing, all others returned to the B-36A designation. The decision to condemn the first B-36A was taken in mid-1946 before the plane's first flight had even taken place. It followed a persuasive presentation by General Twining, who stated that little was known about the B-36's specific construction. Therefore, he thought it prudent to static test one B-36 to destruction. By his reasoning, experience had shown that the Air Force would have been unable to utilize its bombers properly in WWII if it did not have this strategy in place in the past. For example, he claimed that the B-17, initially intended for a gross weight of 37,000 pounds, flew throughout the Second World War at 64,000 pounds. This result could never have been achieved without a precise understanding of the strengths of the bomber's component parts.

General Twining won the argument. On August 30, 1947, 44-92004, just two days after its inaugural flight, it was flown to Wright Air Development Center at Wright-Patterson AFB, Dayton, Ohio, for structural testing. Unfortunately, the plane only ever made these two flights. Later, it was redesignated YB-36A-1-CF, and, as General Twining wanted, the B-36 was disassembled at Wright-Patterson AFB, and each of its parts was subjected to stress tests. Eventually, BM-004 was utterly destroyed.

Meanwhile, the Consolidated Vultee Fort Worth, Texas, Division efficiently produced 100 B-36A bombers for the Air Force. The B-36A was considerably different from the original XB-36, with significant differences in the 4-wheel main gear units and a redesigned nose compartment, including the bubble canopy. The new gear permitted the 160-ton giant bomber to operate from any field capable of sustaining Boeing B-29 operations compared to the three airfields from which the single-wheel XB-36 could operate. The armament of the B-36A included eight remote-controlled turrets mounting twin 20-mm cannon. In addition, there was a possibility for 'parasite' fighters to be carried in the tremendous bomb bay for high-altitude defense measures

On June 26, 1948, SAC received 22 B-36A bombers without gun armament, as their primary purpose was to train the crews. Four days after the first B-36A was received, a demonstration dropping of 72 bombs weighing 1,000 lb. each took place.

The first production type immediately became the B-36A. Soon after, the USAF decided that the first twenty-two aircraft would use Pratt & Whitney R-4360-25 engines, giving 3,000 hp at takeoff. However, the following variant would use the improved Pratt & Whitney R-4360-41, delivering 3,500 hp. with water injection. The latter would then be designated B-36-B. The B-36Bs did not last long. Later, all but three B-36Bs were retrospectively modified to the B-36D standard.

The ill-fated first B-36A, 44-92004, seen here in one of only two flights it ever made before being destroyed in stress tests at the Wright Air Development Center at Wright-Patterson AFB, Dayton, Ohio, in 1947. (T. Gianna, USAF)

B-36A Specifications

TYPE	POWERPLANT	DIMENSIONS WINGSPAN / LENGTH / HEIGHT	WEIGHTS Empty / Combat / Gross / Max. t.o.w.	SPEEDS Cruise / Combat Speed / At 31,600 ft – Basic Range/ Max Speed	SERVICE CEILING / RANGE
B-36A	Pratt & Whitney R-4360-25. 3250 hp takeoff. 3000 hp at 40,000 ft	230 ft/ 162 ft 0.25 in / 46 ft 10 in	135,020 lb./ 212,800 lb. / 310,380 lb. / 311,000 lb.	218 mph / 290 mph / 356 mph / 345 mph	39,500 ft / 9,136 miles
B-36A WEAPONS		Unarmed version			

The B-36B variant had the more powerful Pratt & Whitney R-4360-41 engines with 3500 hp. It was also the first version to be armed. Convair B-36B, 44-92026, was the first in a series of 73 B-36Bs produced, and its first flight occurred on July 8, 1948. The maximum speed of 380 mph and altitude of 34,500 ft. were not very impressive. In fact, it amplified the impotence of propeller giants in the world of much faster and incomparably more agile jet fighter types in the new jet era. As a result, the arch enemy of the B-36, SAC Commander General George C. Kenney, advised abandoning the B-36 project in favor of the quicker, cheaper, and more compact Boeing B-50.

In March 1947, Convair and Pratt & Whitney proposed the R-4360-VDT (Variable Discharge Turbine) powerplant developing 4,300 hp. The manufacturer estimated that with these new engines, the B-36 would have a maximum speed of 410 mph, reach a ceiling of 44,000 ft., and would have a range of 10,000 miles, with a bomb load of 9,900 lb. However, the VDT engine could only be used in traction, which basically meant changing the whole structure of the wings so that the propellers were no longer behind, but in front of the wings. Nevertheless, the B-36C looked promising enough to result in Convair being authorized to proceed with a YB-36C prototype in July 1947. Funding for this project was by cutting three airframes from the existing 100 aircraft contract. However, that was not the final act. General Spaatz, Air Force Chief of Staff, intervened to cancel the YB-36C on August 22, 1947. Convair responded on September 4 with a proposal to install VDT engines on the remaining 34 aircraft, s/n 44-92065 to 44-92098, with the cost covered by a decrease of the initial contract to 95 aircraft.

Convair proposed converting 61 B-36A and B-36B aircraft to the same standard as part of the same plan, which was considered by the Air Force but ultimately rejected. The B-36C project was kept alive, and Convair was permitted to proceed on December 5. However, it quickly became clear that the plan to install the VDT engines was troubled with complications. Nevertheless, the Convair managers continued to push the VDT engine idea but faced constant constraints. The USAF's initial budget had been greatly exceeded, already amounting to 1 million dollars per month in 1947, which was a phenomenal amount at the time.

The USAF was inclined to abandon the B-36 in favor of the B-50 and the new jet bombers. In fact, the Air Force budget submitted to Congress in June 1948 did not include any new B-36s; only Northrop B-49s, Boeing B-50s, and Boeing B-54s were ordered.

Convair was on the razor's edge. Something had to give. Something did, and Erickson was the protagonist once again. To demonstrate the true worth of the B-36, Beryl Erickson and A. S. Witchell took off on a B-36A from Fort Worth. Between April 8 and 9, 1948, the two pilots flew to San Diego, returned to Fort Worth, and did it again three times over. The B-36A covered 6,900 miles in 33 hours and 10 minutes of flight, and it still had fuel to fly for 200 more miles. The aircraft flew under normal operating conditions, with a bomb load of 4,500 lb. dropped halfway through the flight on the Wilcox shooting range in Arizona from an altitude of 25,000 ft. The cruising speed was 215 mph, with two engines stopped in flight due to carburetion problems.

Beryl Erickson wanted to rub salt into the B-36's detractors' wounds with another trial flight. Once again, Erickson and A. S. Witchell accomplished the flight with the same aircraft. This time the pilots flew between Fort Worth, Hawaii nonstop, and back again. This long flight of around 8,050 miles was completed in 33 hours and 8 minutes without the slightest incident. Because of the two Convair pilots, the image of the B-36 was restored in the eyes of many. The pilots' actions ensured the program's survival. Before these two flights, even though the B-36 program was maintained, it was hotly debated until May 28, 1948. That was the day when the speed and maneuverability of the great bomber, still doubted by many, were demonstrated.

Beryl Erickson caught wind of another opportunity to show the world the worth of the B-36. The first Secretary of the Air Force, William Stuart Symington III, took off from Dallas aboard a Douglas C-54 Skymaster. Beryl Erickson and his new partner in crime, W. P. Easley, took off aboard a B-36A. They headed for the Secretary of the Air Force's C-54. Once they were near the Skymaster and at the same speed, Erickson carried out the planned experiment. He put the engines to full power, then quickly outpaced the slower Douglas and began to fly around it. Watching events unfold above them, people on the ground saw the large B-36 circling the C-54 like a vulture around a carcass. Symington could not believe his eyes. From that moment on, he was a genuine fan of the B-36 Peacemaker.

The Air Force decided to conduct some experiments to compare the performance of the basic B-36 with that of other bombers under comparable conditions. The new yardsticks assessed each bomber's four most critical and interdependent characteristics: speed, range, altitude, and load capacity.

The basic B-36 performed better in tests, though not as well as Convair hoped. The findings demonstrated that the slow B-36 outperformed the B-50 in long-range cruising speed and had a higher altitude, a bigger cargo capacity, and a broader battle radius than the B-50 or the B-54, which was a B-50 derivative. It was examined but abandoned in 1949. Now, it appeared that the B-36 would end up being far better than expected. If that were the case, any premature contract termination might ruin the program just as it was about to succeed. The Air Force was spared more uncertainty when the Russian blockade of West Berlin started its Berlin blockade on June 18, 1948. Deeply alarmed by the Soviet Union's assertiveness, on June 25, Air Force Secretary W. Stuart Symington and other top USAF officials unanimously decided to stick with the B-36.

The survival of the B-36 program was now assured and not a moment too soon. With the USSR's blockade of Berlin a few days earlier, East-West relations went very sour. The B-36 had suddenly become a top strategic asset. Even the Air Force personnel that were the most hostile to the B-36, General Kenney and General Norstad, agreed that when faced with such a delicate international situation, the large Convair bomber was the only aircraft in its category that could take care of the situation. The Boeing B-50 and Northrop B-49 suddenly drifted into insignificance; the new master was now the 'Peacemaker.'

Thus as mentioned previously, the first five Convair B-36As were delivered to SAC's 7th Bomb Wing at Carswell AFB in Texas. These and the subsequent 17 B-36A aircraft were unarmed and primarily utilized for crew training. Consequently, they were reconfigured to the reconnaissance configuration before they joined operational B-36 units.

The more powerful Pratt & Whitney R-4360-41 engines were made available with the B-36B variant. The extra power greatly enhanced the bomber's performance. With this version, SAC started to test the B-36 to the extent of its capabilities. On May 28, 1948, a B-36 completed a 36-hour flight that covered more than 8,000 miles in a circuitous route around the United States. Then, on the eighth anniversary of the Japanese attack on Pearl Harbor, December 7, 1948, a B-36A took off from its base at Carswell, Fort Worth AFB, to drop 10,000 lb. of bombs near Oahu. The B-36 then returned to Fort Worth after flying nonstop for 8,000 miles at an average speed of 315 mph without being detected by the island's air defense.

Finally, 22 B-36As were produced, while the total B-36Bs manufactured was 62. In fact, Convair manufactured 78 B-36Bs, but the Air Force had 11 of them modified before formal acceptance. Four of the eleven were designated as B-36Ds, while the other seven were designated as RB-36Ds.

B-36B Specifications

TYPE	POWERPLANT	DIMENSIONS WINGSPAN / LENGTH / HEIGHT	WEIGHTS Empty / Combat / Gross / Max. t.o.w.	SPEEDS Cruise / At 34,500 ft Basic Mission/ Max Speed	SERVICE CEILING / RANGE
B-36B	Pratt & Whitney R-4360-41, 3500 hp	230 ft/ 162 ft 0.25 in / 46 ft 10 in	140,640 lb./ 227,700 lb / 328,000 lb. / 311,000 lb	203 mph / 380 mph / 381 mph	42,500 ft / 8,820 miles
B-36B WEAPONS		Two 20-mm M24A1 cannon each in six retractable, remotely-controlled fuselage turrets. Tail turret, and nose mounting, with 92,000 rounds of ammunition. Normal bomb load up to 72,000 lb. Maximum bomb load 86,000 lb.			

A B-36B- 1-CF, 44-92033, which has six 3,500 hp. P&W R-4360-41 engines. It also has full armament installed, comprising 16 20 mm cannon in 8 barbettes – six in the fuselage, one in the tail, and two in the nose. 73 of these variants were manufactured. (USAF)

This is another photo of B-36B- 1-CF, 44-92033, surrounded by a sky full of clouds. (T. Gianna, USAF)

B-36 CREW TRAINING SYSTEM

Equipment especially suited for the training of the future crews of the Convair B-36A was created by the Consolidated Vultee Corporation in liaison with Air Training Command and sent by air to the Carswell Air Force Base, located northwest of Fort Worth, Texas. It was the first base to receive the Convair B-36A.

The 25 flat panels had hydraulic and electrical components or tools that assured their control. With these enormous boards, the trainees could get familiarized with the Convair B-36 and get on the aircraft sooner.

The B-36A aircrews went through their training smoothly and thoroughly. They learned the drills they would carry out, whether on the ground or in flight. They were taught by fifteen highly specialized instructors and directed by two officers. The training method was not new; during World War II, Consolidated-Vultee produced similar panels for the training of pilots, mechanics, radio, and machine operators of the B-24 Liberator.

The practical exercises, which the pilots completed, were also accompanied by theoretical courses. They were created by the manufacturers of the plane and its equipment. Companies like Pratt & Whitney, Curtiss-Wright Corporation, Minneapolis-Honeywell Regulator Company, and the Chanute and Boca Raton schools of the Air Force prepared detailed manuals on the installation, operation, adjustment, maintenance, etc., of all the devices that were installed on the Convair B-36As.

The 25 instruction boards for the practical side of the training included the levers and the instruments related to the synchronization of the propellers, the distribution of the loads in the plane, the opening of the bomb hold and release of the bombs, the engine cooling and lubrication system, the turbo-compressor and autopilot adjustment, various weapons-related devices, radio, and radar, among others. In addition, the equipment on the boards was electrically driven, so the students could practice as if they were actually on board the aircraft.

The electrical components of the Convair B-36A were supplied by two generators, each powered by a 145 HP Continental engine mounted on a four-wheel drive system. The drive wire was 30 meters long to keep the power units away from the panels and minimize any negative electrical issues.

These drive panels were designed to operate in any extreme climate, hot or cold. They were housed in cases whose weight and volume was calculated so they could be quickly shipped on cargo planes for rapid transport from one base to another. Metal slides in the bottom of the boxes allowed the panels to be rapidly moved in or out. Finally, after all, was said and done, the crews were ready to fly the 'Big Stick.'

THE USAF BUYS MORE B-36 BOMBERS

The year 1949 was looking good for Convair. The USAF announced it would buy more B-36 bombers early that year. However, that decision came at a high price. In order to obtain the money to buy the additional B-36 aircraft, the USAF, in March, canceled the following contracts: 30 Northrop RB-49 Flying Wing bombers, 51 North American B-45 four-engine bombers, 118 North American F-93A (experimental) jet fighters, 30 Northrop C-125 Raider assault transports, and 10 Kellett H-10 (experimental) helicopters. By canceling orders for 239 aircraft worth $300 million, the USAF eventually had enough money to purchase just 39 Convair B-36s. This statistic alone shows the tremendous cost of strategic airpower in the 1950s.

Many experts predicted that the Air Force would cancel the B-36 order. However, that was no longer the case. After the successful long-range flights and bomb drops orchestrated by Convair and its pilots, the USAF was seduced to purchase the B-36 bombers over the previous months.

Washington was buzzing with rumors about why the B-36 decision was made, but in essence, it merely marked a significant shift in the USAF's focus on strategic airpower. Additionally, the B-36 was the only bomber in the world that could supposedly travel 10,000 miles while carrying a load of 10,000 lb. bombs. Despite its shortcomings, the B-36 was the longest-range aircraft in production at the time, and if a world war were to break out anytime in the near future, then the USAF's B-36s would be up to the fight.

Subsequently, the USAF was given $300 million, of which $200 million was for brand-new Convair B-36 bombers, and $100 million was for modifications and 'improvements' to already-existing B-36 bombers.

In January 1949, Convair wanted to put on another show with its B-36. A Convair B-36B dropped two 20-ton bombs 1,200 miles from its takeoff location, the first from 35,000 ft. and the second from nearly 40,000 ft. The number of bombs it carried was the most by a single aircraft. According to the USAF Chief of Staff, General Hoyt S. Vandenberg, the B-36s, flying at 40,000 ft., could not be intercepted by any American jet fighters in the Air Force at the time. Moreover, he stated experience showed that if a jet fighter attacked the B-36 from behind, the fighter's engine was liable to blow up in the turbulent slipstream from the bomber's six giant engines. In light of this information, the USAF canceled contracts for other bombers and fighters in favor of more Convair B-36s.

The world's largest bomber, which progressed in one short year from a plane whose production contract cancellation was ready to be signed, to the number one strategic weapon of the United States. In another display of its potency, the Air Force incredulously stated the colossal bomber completed a 9600-mile nonstop flight, during which it dropped a 10,000 lb. bomb load and two 42,000 lb. bombs in a single flight, reached an altitude of 46,000 ft. and remained above 40,000 ft. for 12 hours.

WAS THE NAVY RIGHT ABOUT THE B-36?

Maybe, the Navy was right about its assertions regarding the B-36. The much-criticized bomber was not the epitome of perfection. The Navy, on one corner, slammed the Air Force for its preference for a slow piston engine bomber that was incapable of achieving the 10,000 lb. 10,000-mile targets. The Air Force in the other corner was as persistent as ever. It claimed the B-36 was the best option for America in the new Cold War era. But, one could argue the B-36 was far from being the bomber that the Air Force had supported since its inception in 1941.

After spending so much money on their favorite aircraft, the USAF thought the 10,000-mile target was plainly in sight. Unfortunately, it appeared that the Air Force's top planners were unwilling to cut their losses and embark on a new project to achieve the Air Force's aims. The dreaded 10,000-mile figure was not easily attainable for the B-36. That was a fact that the Air Force could not ignore.

For example, by pushing up the B-36's fuel capacity, the military load had to be reduced, while the bomber's airframe became larger and slower. One glaring fact was that the B-36 was forced to carry more fuel than its own bomb load. For instance, when comparing fuel increases, the XB-36 had to hold 1,200 US gallons of fuel, weighing around 9,000 lb., whereas the B-36B was burdened with at least 11,000 lb.

Regarding Convair, the company received only a few modest contracts from the critical 1949 70-Group rearmament program. Then, a few months later, the Air Force suddenly placed orders worth more than $300 million with Convair. The individual who accomplished this incredible deal was Consolidated Vultee's boss, millionaire Floyd Odium, an American lawyer and entrepreneur who bought out Convair almost two years earlier in a down market.

THE B-36 AND ITS LIMITS

Claims by the USAF officials about the B-36B having a proven service ceiling of more than 40,000 ft. with a full combat load were hard to swallow by some detractors of the bomber. As mentioned earlier, some Air Force officials even claimed that the service ceiling was as high as 46,000 ft. These assertions, however, were dependent on the aircraft's weight definition. The B-36B was neither a speedster nor a streamlined plane. With its actual initial gross weight of 326,000 lb., it was aerodynamically incapable of attaining a service ceiling of more than 28,000 ft. In the instance of a bomber with a 5,000-mile radius, the takeoff weight condition was not a reliable indicator of combat performance; therefore, let's look at the situation closer to the target area instead.

At the absolute target radius of 5,000 miles, after consuming just over 60% of the fuel load and immediately before the 10,000 lb. bomb load is released, the gross weight drops to about 216,000 lb. As a result, the most optimistic evaluation predicted that the B-36B's maximum service ceiling would be around 39,000 ft. When looking at the data, it was assumed that the Wasp Major R-4360-41 engines, which were installed on the B-36B at the time, would give an average continuous output of 2,650 b.h.p up to a critical supercharger height of 35,000 ft. This drag coefficient was based on the retraction of each gun turret.

Let's say the B-36B could have reached a height close to 40,000 ft. in the target region. That would only be an escape ceiling rather than a combat altitude since the maneuverability of a 160-ton bomber at its service ceiling had to be of short duration. The B-36's maximum speeds showed the Big Stick's initial speed to be 290 mph at 20,000 ft. with a starting weight of 326,000 lb. Then, the rate rose to approximately 360 mph at 35,000 ft., as the target weight reached 216,000 lb. However, if the aircraft was pushed to its service ceiling, its maximum speed would drop to about 340 mph. Therefore, the optimum fighting altitude for the B-36 was about 35,000 ft.

On the other hand, a long-range bomber had to spend almost all of its time cruising at its optimum-range speed, or it would not be able to reach its specified range. Regrettably, this is where the piston-engine bomber failed because its optimal cruising speed was well below the maximum. This situation occurs because, aerodynamically, the aircraft has to fly at its highest lift/drag ratio. At the same time, the engine should run at its most efficient speed to have the lowest specific fuel consumption. Thus, the average cruising speed of the B-36 on all its long-distance flights was only around 230 mph. That was a little lower than the cruising system appeared to support. However, as the cruising technique was refined, it would be possible to increase the speed to around 250 mph.

In the other dubious statement, the USAF made an extraordinary admission. It announced its jet fighters could not deal with the B-36 at its maximum bombing altitude, which was reported to be at least 40,000 ft. The claim was somewhat misleading. Firstly, the US fighters probably could not attain the high altitude of the B-36 due to heavy wing loading, a lack of engine power at high altitudes, or a mix of the two. Moreover, the USAF based that claim on the theory that the 'enemy' jet-fighter designs in the 1950s had followed American practice. In the early 1950s, American jets were high-speed and high-wing-loading fighters. That was not the case in Europe. The European fighters were built with a lighter wing-loading, giving them fast climb for rapid interception and agile maneuverability at high altitudes. Some European fighter planes could fly to 40,000 ft. in approximately 8 minutes and continue to combat from there up to well over 50,000 ft. In an actual war, the USAF's fighters would not be chasing the B-36 but enemy fighters with higher maneuverability. Thus the USAF's assertion did not mean much in this situation.

Incidentally, the B-36B was so heavily overloaded at the start of a mission that it couldn't fly in its minimum-drag attitude until it had burned up part of its fuel and oil.

Nonetheless, given that the B-36B flew one 9,600-mile mission with a 10,000 lb. bomb load on the outward run, it was able to meet its original specification range of 10,000 miles, although it required an additional 6,000 gallons of fuel. Under ideal conditions, an absolute maximum range of 10,600 miles while carrying 27,000 US gallons of fuel was achievable.

The B-36 served as yet another reminder to American taxpayers that the USAF was throwing caution to the wind in pursuit of more taxpayer money. Still, the B-36 orders kept piling up. By the middle of 1949, the number of B-36s ordered had risen to 170, with around 60 already being delivered. The USAF intended to standardize the B-36 as their long-range strategic bomber, to the exclusion of everything else on the horizon. The Air Force's senior military commanders, who became so emotionally invested in the B-36 project, lacked the courage to abandon it. It looked like the Navy may have been right about the B-36 and its fans in the Air Force. What saved the day for the Air Force was the following B-36 variant, the B-36D.

B-36D

Following the Second World War, the Air Force had piston-engine bombers such as the B-29 and B-50 at its disposal. The Convair B-36 then appeared on the scene in the late 1940s. However, all three bombers were ineffective because jet fighter interceptors were appearing practically everywhere in the world, flying faster and higher and with greater agility. In the case of another war, an opposing jet fighter interceptor would easily pick off the slow-flying prop bombers.

Convair was aware of the fighter problem in October 1948 and advocated adding turbojet engines to enhance the B-36's engine power. Following that, Convair upgraded one of the B-36Bs, 44-92057, to a test version, YB-36D. Four Allison J35 jet-assist engines were installed in pods beneath the wings on the outer ends as part of the modification. However, this was just a temporary remedy, as the production B-36D's future powerplants would be two pairs of J47-GE-19 turbojets. On March 26, 1949, the upgraded YB-36D flew for the first time.

Field performance was considerably enhanced, with the jet-assisted YB-36D reaching speeds above 400 mph at 35,000 ft. and a maximum altitude of 40,000 ft. On the downside, the combat radius for the B-36B was lowered from 4,300 miles to 3,530 miles. At the same time, the overall mission duration was cut from over 42 hours to 31.5 hours.

Just over three months later, on July 11, 1949, the first actual production, B-36D, took off with the General Electric J47-GE-19 turbojets for the first time. In performance, the J47-powered B-36D achieved an over-the-target speed of 435 mph at 50,000 ft. Meanwhile, the 'D' variant had many new features installed, including a K-3A bombing and navigation system compared to the B-36B's basic APG-24 radar. The K-3A system enabled a single crew member to operate as a radar operator and bombardier. Furthermore, instead of the B-36B's APG-3 radar, the B-36D had the AN/APG-32 radar to control the tail turret.

Additionally, the B-36D's takeoff weight increased from 330,000 lb. to 370,000 lb., and its landing weight climbed from 328,000 lb. to 357,000 lb. Finally, the aircraft was outfitted with innovative snap-action bomb-bay doors that opened and closed in 2 seconds flat, compared to the slower sliding bomb-bay doors of the previous B-36As and B-36Bs. After the success of the B-36D's first flight, the USAF took notice but did not accept any of the B-36Ds immediately; it would take another year before any joined the Air Force inventory.

The first flight of the YB-36D, March 26, 1949. This aircraft was originally a B-36B, 44-92057. (T. Gianna, USAF)

B-36, 44-92096, number 93, was a B36B-15-CF, of which 34 were built. It was accepted by USAF in July 1950. This aircraft was used for propeller vibration tests. It was assigned to 95th Bombardment Wing, at Biggs AFB, El Paso, Texas. (T. Gianna, USAF)

This is a photo of the B-36D, 44-92057, Number 54, in flight. It was a Featherweight II, and It served with 325th Bombardment Squadron, 92nd Bombardment Wing, Fairchild AFB, Spokane, Washington. It suffered structural failure at Fairchild AFB on March 31, 1954. On December 6, 1956, sent to MASDC and scrapped. (T. Gianna, USAF)

This is the same aircraft as in the photo above. Early on, before joining the 325th Bombardment Squadron, it was used for armament tests, and flight tests on the J35 and J47 engines. It also performed four-blade propeller vibration tests, and Pratt & Whitney R-4360-53 engine tests. It was accepted by USAF in January 1951. Convair later converted it to a B-36D-40-CF at its San Diego plant in California. (USAF)

B-36D-5-CF, 49-2652, of the 7th Bombardment Wing, Heavy, Carswell Air Force Base Texas. The 7th BW,H, began converting to B–36 bombers in Jun 1948. It operated two B–36 groups, from December 1948 to February 1951, and three B–36 squadrons from February 1951 to May 1958. (T. Gianna, USAF)

This is a Convair RB-36D of the 28th Strategic Reconnaissance Wing, based at Ellsworth AFB, South Dakota. The 28th SRW, operated B-36s 1949 to 1950, and RB–36s 1950 to 1957. (T. Gianna, USAF)

This is an RB-36D-10-CF, 49-2689, which was accepted by the USAF in January 1951. It was the Strategic reconnaissance model which had two of its four bomb bays replaced by a 14-camera installation. It also had 22 crew members, as opposed to the usual 15. It served with the 28th Strategic Reconnaissance Wing, Rapid City AFB, South Dakota (see previous photo). This wing operated RB-36s 1950-1957. This RB-36D-10-CF was later transferred to the 99th Strategic Reconnaissance Wing, at Fairchild AFB, Spokane, Washington State. The 99th SRW was equipped with RB–36 aircraft at the beginning of 1954, and performed worldwide photographic, electronic, and visual day and night strategic reconnaissance as its primary mission until late 1954. Then, from late 1954 until September 1956 it operated photo reconnaissance only as a secondary mission. The 99th SRW continued to use RB–36s until August 1956. It replaced them with B–52s from December 1956. (T. Gianna, USAF)

An RB-36D-10-CF, 44-92688. It is the same model as the photo above. It was also based with the 28th Strategic Reconnaissance Wing, Rapid City AFB, South Dakota. (T. Gianna, USAF)

A Convair B-36D in flight. (T. Gianna, USAF)

A Convair B-36 in flight near Edwards Air Force Base, California. (National Archives)

THE 'FEATHERWEIGHT PROGRAM'

The USAF began a program in 1954 to boost the abilities of the B-36s. The 'Featherweight program' was launched to lower the weight of all B-36s in the Air Force's inventory. All but the two radar-controlled tail guns in the tail were removed as part of the program. The elimination of six turrets (containing two guns each) also necessitated the withdrawal of six gunners from each aircraft. As a result, the crew number on the B-36 was reduced from 9 to 3. While on the RB-36s, the crew was reduced from 15 to 9. Guns on post-World War II bombers were rendered obsolete in the 1950s by the introduction of more effective anti-bomber weapons, such as missiles that could target a bomber from afar. The 'Featherweight Program' enabled the B-36 bombers to perform better in speed and altitude due to the drop in the aircraft's weight. Moreover, the Strategic Air Command saw the gunnery and defensive armament as the B-36's weakest link. Therefore, SAC was satisfied with their removal.

There were three levels in the 'Featherweight Program.' 'Featherweight I' had all defensive equipment removed from the B-36s except for the two gun turrets. In 'Featherweight II,' the rear compartment crew comfort features, and all the gear supporting the McDonnell XF-85 Goblin parasite fighter, were removed. Lastly, 'Featherweight III' included all the requirements of Featherweight I & II. Eventually, all B-36s were converted to Featherweights.

THE GENERAL ELECTRIC J47 TURBOJET ENGINE

This is a General Electric J47 turbojet engine, similar to the type that the B-36s used. It is a J47-GE-27 engine offering 5,970 pounds-force (27 kN). The B-36's were equipped with the slightly less powerful J47-GE-19 which gave the bomber 5,200 pounds-force (23 kN). (Photo by Hunini)

The J47 engine was a refined version of General Electric Corporation's first axial engine, the J35, which Allison received in 1946. The engine could be fitted in any aircraft intended for the earlier TG-180 (J35), such as the world's fastest fighter, at the time, F-86A Sabre. Other J47A installations included the B-45C Tornado bomber, B-47A Stratojet production versions, the North American Aviation YF-93 Sabre, and the bomber variant of the B-49 Flying Wing. The engine was also specified for the twin-jet underslung pods of the B-36D bombers. The B-36s would eventually receive the later version of the engine, the TG-190C, J47-GE-19, which provided 5,200 lbs. thrust.

In December 1949, General Electric Corp. installed four J47 turbojets on Consolidated Vultee's prototype B-36D, which underwent flight tests at Fort Worth. The four jets were mounted in pairs underneath each outer wing panel and added more than 20,000 lbs. thrust to the 21,000 hp. of the giant bomber's six 3500-hp Pratt & Whitney Wasp Majors, supplying extra power for takeoff and higher speeds within target areas. The Jet pods were finally added to every B-36; those delivered and those under contract.

B-36D 'Featherweight III' Specifications

TYPE	POWERPLANT	DIMENSIONS WINGSPAN / LENGTH / HEIGHT	WEIGHTS Empty / Combat / Gross / Max. t.o.w.	SPEEDS Cruise / At 37,300 ft. / Combat / Max speed	SERVICE CEILING / RANGE
B-36D	Pratt & Whitney R-4360-41, 3500 hp. & General Electric J47-GE-19, 5200 lbs. thrust	230 ft/ 162 ft 0.25 in / 46 ft 10 in	161,371 lb. / 250,300 lb. / 370,000 lb. / 370,000 lb.	220 mph / 417 mph / 413 mph / 406 mph	33,400 ft / 9,436 miles
B-36D WEAPONS		2 x 20 mm (M-24A1) tail guns with 1,200 rounds (600 rounds each). A 72,000 lb. bomb load could be carried.			

B-36 BOMBER OPERATIONS 1950-1951

The first B-36Ds accepted by the Air Force in August 1950 were sent to Eglin AFB for testing, but some of the new aircraft arrived considerably later. In December, SAC operated the following Convair bombers: 38 B-36s, six B-36Ds, and approximately 24 B-36Bs that were later upgraded to the 'D' variant. In addition, units of the Eighth Air Force's 7th Bombardment Wing were equipped with the bombers.

The Air Force announced in January 1950 that extra Convair B-36 bombers would be delivered, 34 in total. This boosted the overall number of orders to more than 200. It was undoubtedly the most powerful long-range air-striking arm in history. In addition, the 200 Peacemakers provided enough aircraft for four heavy bombardment groups, two long-range strategic reconnaissance groups, and plenty of spare aircraft for replacement and special tests.

In August 1950, two months after the beginning of the Korean War, the Navy was the master of the seas around Korea. During that war, several of the Navy's points in the B-36 row from 1949 were validated. There was no doubt that the US Navy could approach the Korean coast with its carriers and carrier aircraft for extended periods.

The new Convair RB-36 (reconnaissance variant) looked like a standard bomber, but its number 1 bomb bay was crammed full of 14 different cameras, one of which had a massive 48-inch focal length lens. In addition, the Air Force was satisfied with the operations of its RB-36s and purchased 67 RB-36E photo-reconnaissance versions to outfit two photo-reconnaissance groups. The ultimate goal of the Air Force was for four B-36 bomber groups and two RB-36 photo-reconnaissance groups.

The B-36 bombers did not leave the skies of the United States until 1951. That year, six B-36Ds made the first official overseas flight from Carswell AFB to Great Britain on January 16, 1950. The six 'Big Sticks' landed at Lakenheath RAF Base and returned to Carswell on January 20. A similar journey to French Morocco was accomplished on December 3, 1950, when another six B-36 bombers from the 11th Bombardment Wing arrived in Sidi Slimane, a small city in the northwestern center of Morocco, after flying nonstop from Carswell.

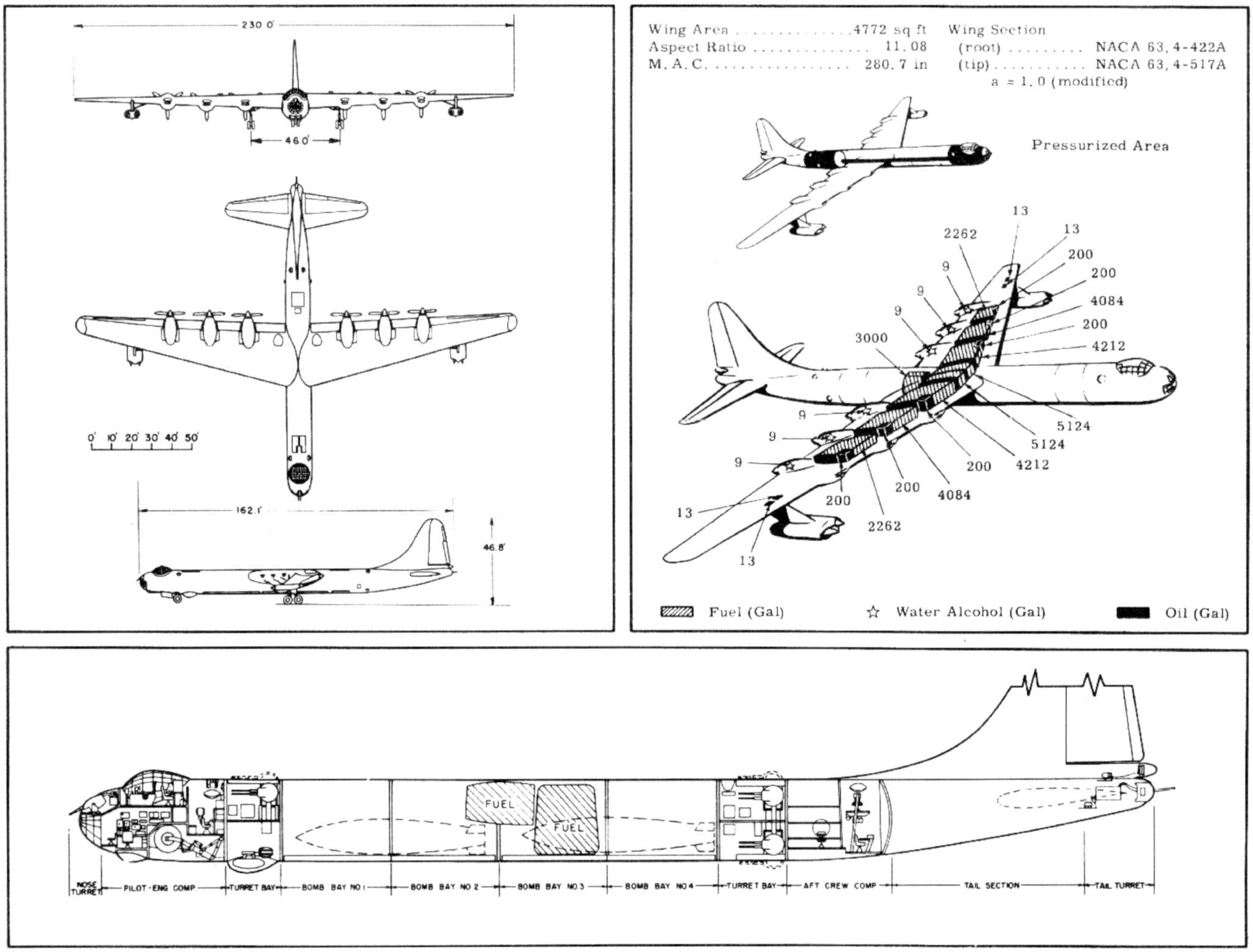

CONVAIR B-36D 'PEACEMAKER' - STANDARD AIRCRAFT DIMENSIONS & CHARACTERISTICS (Graphics: USAF)

THE FORTUNATE B-36

On January 26, 1952, a B-36D made an incredible undershot landing on the outskirts of Salisbury near Boscombe Down, the home of a British military aircraft testing site in the United Kingdom. The pilot, who was returning to the air base after a training trip from America, came down through low clouds and confused the approach lights for the runway lights. As a result, he landed several hundred ft. short of the runway, and the colossal B-36 bomber rolled over a haystack and the main road before coming to a rest. Then, miraculously, the pilot is said to have called ATC for taxiing!

Earlier, the locals were horrified to hear the B-36D flying low overhead shortly before midnight, seemingly in danger or having lost its course. The bomber hovered low over the city for two hours, rattling windows and doors and filling the air with a harsh, growling noise from its six Pratt & Whitney R-4360 engines. It eventually flew northward, narrowly avoiding a Hotel near the city center, and landed in the vast snow-covered field.

High-ranking United States Third Tactical Air Force officers arrived on the site and observed specially-drafted crash teams in the dozens working in gale-force winds and rain to move the big bomber from the side of the road where it had stopped. Moreover, armed US sentries formed a large ring around the highly secretive nuclear bomber.

The aircraft's bogie undercarriage appeared to have survived the dramatic landing. After some repairs, the B-36 took off from Boscombe Down on its return flight to Carswell Air Force Base, Fort Worth, Texas. The only apparent damage was the piercing of two small holes in the nose of the 'Big Stick,' possibly by fencing posts or flying stones. Lucky pilots? Or lucky B-36?

BERYL ERICKSON - FLYING THE B-36D

Beryl Erickson, the pilot who flew the XB-36 prototype, was pleased with himself. Aside from flying the prototype, he was also the pilot with the most flying hours on the B-36; despite the aircraft's inherent complexity, Erickson concluded, after more than 1,000 flight hours, that the B-36 was as easy to pilot as the most good-natured bomber. However, Erickson did not state which bomber that was. He also said it was incorrect to state that the difficulty of flying increased with the size of the plane.

The only challenging flight for Erickson was when he flew the prototype on August 8, 1946. At that time, he had a lot of weight on his shoulders. He had worked on the B-36 project for over two years and knew the plane had cost the nation a lot. It also had many detractors.

His training to be a B-36 pilot has been relatively easy. He said it took a four-engine pilot less time to switch to the B-36 than it did from a single-engine to the original four-engine aircraft. In addition, Fort Worth pilots were occasionally asked to fly a few hours on the Stinson Voyager (a light utility monoplane of the 1940s) to maintain their original flying skills and to be able to adapt easier.

The best advice he gave the new B-36 pilots was to be constantly aware of the aircraft's weight and dimensions. On landing, for example, the pilot had to think, among other things, of the distance of the landing gear below and behind his position; otherwise, the pilot risked hitting obstacles.

As soon as the pilot developed this sense and understood the mechanical characteristics of the B-36, he could fly the massive bomber as casually as any other plane.

The pilots appreciated the safety of flying with ten engines. The loss of power from one, two, or even four engines would be minimal. Similarly, if the B-36 were flying over the United States in severe weather, the pilot would not have to worry about alternate airfields because he had enough fuel and engines to fly to a suitable airport, even if it was on the opposite side of the country!

The B 36's primary crew consisted of two pilots, two flight engineers, and two observers. The pilots and flight engineers collaborated closely and formed what was the traditional concept of crew during that era. They were stationed in the cockpit, near the front, and at the station's highest point.

The copilot controlled the four J-47 turbojet engines, while the two flight engineers were in charge of the six-piston engines and the aircraft's hydraulic and electrical systems. Meanwhile, the crew at the back of the plane kept an eye on the engines, landing gear, and flaps.

Two to three flight test engineers served as additional test personnel and were responsible for operating the instruments and specifically mounted recorders. In addition, they performed comparable chores like recording fuel information and inspecting test equipment. The Air Force's B-36s would often carry large crews. In addition to the primary crew members, photographers, radar operators, navigators, and gunners, were also included. At the very least, the B-36 was large enough to accommodate all of them.

A TYPICAL B-36 HIGH-ALTITUDE FLIGHT

On any high-altitude flight on the B-36 bomber, the preflight inspection was the first item on the agenda for the crew. Then, depending on the crew members' responsibilities, each checked the aircraft to ensure that the fuel and oil supplies were adequate for the estimated flight time and that everything was in working order. Before taking his seat on the plane, the pilot had already mapped out each crew member's role for the duration of the flight.

The crew members provided the pilot with their initial evaluation of the aircraft's readiness for flight. Next, the pilot conducted the preflight inspection while the flight engineer, seated in front of his panel, directly behind the pilot, started the aircraft's six-piston engines. While the engines were being warmed up, the pilot continued to go through the checklist.

One crew member, yet to board the aircraft, watched the engines from the ground while communicating with the pilot via a telephone linked to a hull socket. He, too, had a checklist, and before boarding, he told the pilot that the final external examination of the aircraft had been performed and that it was fully ready for takeoff.

The safety locks and protective coverings were removed, the armament doors were closed, and the grounding wires were detached from the aircraft. When everything was ready, the crew member gave the OK to the pilot and unplugged his phone. With everyone on board, the pilot contacted ATC and, with their consent, taxied to his takeoff position.

While the B-36 was taxiing, observers scanned for obstacles from the back of the cabin. These individuals served as the pilot's eyes and ears. In preparation for takeoff, the two flight engineers ran their lengthy checklist during the roll as the copilot in command started the four General Electric J-47-GE-19 jet engines, which did not require heating like the six-piston engines. However, due to their high fuel consumption and lack of need for warming up, they were only started during takeoff.

With the Big Stick's ten engines on idle, the total thrust was such that the propellers of the four inside engines were set to reverse to avoid excessive usage of the brakes in the final phase of rolling. At the end of the runway, the pilot spun around the B-36, adjusted the trim tabs to the takeoff position, returned the propellers to the proper pitch, and only unlocked his primary flight controls. Furthermore, these were kept in check to reduce potential damage from gusts. They were unlocked by lowering a simple electric switch, which controlled the hydraulic release system.

While the engines were brought up to takeoff power, the aircraft was held back by its brakes. Engine cooling was unnecessary because the B-36 had forced cooling fans that delivered airflow even during the run-up. Then the airplane accelerated after the brakes were released. The pilot relied on the very efficient rudder at this speed. He pulled back on the stick to lift the nose gear off the runway. The aircraft seemed to be already in flight to the occupants sitting in the front seat, but it was still rolling. Instead, it had taken on a nose-up attitude with the pilot's cabin and progressively rose to a height roughly four stories high. The takeoff speed approached, and the giant bomber took off. The pilot used the indicated airspeed to control the aircraft's pitch attitude. The copilot retracted the gear and flaps on the pilot's command by moving electric controls. Despite the weight of the takeoff, the quality of the ascent was superb.

In the older World War II heavy bombers, a takeoff at full load produced certain anxiety for the crew because of a possible engine failure. However, with the B-36 and its tremendous power reserve, the loss of engine speed was not an issue. After the B-36 completed the takeoff and reached an altitude of 3,200 ft., the engines' power was lowered to maximum cruise level, which was employed for the ascent until the aircraft reached its service ceiling. During this time, the flight of the B-36 was a matter of correcting and adjusting engine speeds as required, with the copilot taking care of the J47 turbojets while the flight engineer tended to the Pratt & Whitney R-4360 piston engines. The second flight engineer was tasked with maintaining the 160,000-watt electric network, while the pilot was focused on attaining the right flying angle while keeping a consistent rate of ascent. He constantly passed information to the flight engineers from whom he received remarks.

Regardless of the size of the flaps to be operated, the springs-tabs were sufficiently efficient. The B-36's flying procedure was very standard and required minimal pilot effort. However, during the flight's progress, the pilot used the elevator trim tabs to assist the action of the spring tabs. When the B-36 was in level flight, the distribution of fuel and engines along the wing span enabled the aircraft to handle gusts effectively. However, this weight distribution was paid for by lateral inertia (when an aircraft attempts to roll at high speed, centrifugal forces cause it to buck, pitch, and yaw about all three axes), which the pilot had to be aware of. Before the rolling motion had concluded, he had to prepare a recovery maneuver, stop, and even counter with the ailerons. A few degrees per second was the rate at which the B-36 rotated about its longitudinal axis (rolling movement). The B-36's linear speed (the rate at which an aircraft travels in a straight line) at the wingtip was comparable to that of a small plane. By virtue of its wing span, the B-36 required a great distance to execute a banking turn.

The pilot had to avoid giving the controls too much leeway and meticulously plan the recovery procedure from start to finish.

When the airplane began to climb, the wing, tail, and propeller thermal deicers were activated. The pilot operated the flight controls while correcting the effects of cloud turbulence. The forward and aft cabins were pressurized after passing 8,000 ft.

During the ascent into the stratosphere, the flight engineer increased the speed of the engine cooling fans. Level flight was attained when the specified service ceiling of 40,000 ft. was reached. Then, the autopilot was activated, and the pilot asked about the engine's performance. The B-36 got shaken violently during full-speed flight by clear areas of turbulence that one would not expect to experience at these altitudes. As a result, the B-36's fuselage flexed from time to time.

The extremely long B-36 flights, both in distance and time, posed a never-before encountered problem of crew fatigue. With four relief crew members and six rest bunks, Convair improved this problem. Convair outfitted the B-36D models with various home-like comforts, which put the aircrew in the finest physical and mental states possible. They included a kitchen with two electric ovens, a fridge, and pantries. Crew members could eat on folding tables and chairs instead of on the go. Two new bunks were installed in the forward cabin; each bunk, including the six, already installed, had a mattress, bolster, and blankets. Two sinks in the fore and aft of the B-36D completed the layout. Furthermore, slots for personal equipment were situated throughout the fuselage.

With the series of test flights over, the pilot resumed his place, disconnected the autopilot, and prepared for the descent. The copilot switched off the J47 engines while the flight engineer shut down the two outer piston engines, whose propellers he feathered. Fuel consumption reduced the weight of the airplane, which required less power. As a result, the descent was rapid and took place at the level of dive speed or steep descent (the absolute maximum speed above which the aircraft must not fly).

As the bomber descended to less than 10,000 ft., the pressurization in both cabins was cut. The speed was reduced by a temporary leveling off. Then, the landing gear was lowered, and one of the observers left his rear compartment to check and confirm that the landing gear was securely locked, which is why it was lowered so early.

With only four engines running, the B-36 reached the approach area at a slow speed. Lightened by its half-extended flaps, the B-36 gave the impression of floating in the air. During the final approach, the flaps stayed extended. The engine speed was increased in anticipation of clearance from the control tower. If the weather was fine, the pilot would visually inspect his descent gradient but still refer to his flight instruments.

The airplane approached the runway at a height that may appear excessive to others but not the pilots. When the nine primary wheels of the landing gear touched down, the front wheels touched down in turn. To lessen thrust, the pilot asked for reverse pitching of the four propellers. The flight engineer complied by moving his joysticks to the desired position while the pilot locked his control surfaces and guided the aircraft to a halt using the nose wheel and brakes. After taxiing, the flight engineer shut off the two outer engines, leaving only the two inner engines running. After stopping the airplane, all that was left to do was turn off the remaining two engines and report to the office for a post-flight briefing. A thorough flight report had to be written and reviewed by Air Force officials. After all that, the crew could get a well-earned rest.

B-36D, ship number 54, banks to the right, as it heads for the stratosphere. (T. Gianna, USAF)

THE RB-36 'D' AND 'E' VARIANTS

After the B-36D, the Air Force expressed interest in purchasing the RB-36D. That was the photo-reconnaissance model of the 'Big Stick,' which featured 14 cameras and electronics for the aircraft's primary missions. The cameras were housed in two of the B-36's bomb chambers. Remarkably, the lens on one of the cameras was enormous, measuring 48 inches. The interesting thing is the RB-36D made its first flight less than six months after the first B-36D bomber version made its first flight. Six months later, the Air Force began receiving some reconnaissance models.

The RB-36's missions included all-purpose strategic reconnaissance, charting, day and night mapping, and determining bomb damage. Furthermore, the RB-36D also carried 22 crew members, seven more than the B-36D's 15. Convair produced a total of 17 RB-36Ds for the Air Force.

The first photo-reconnaissance Peacemaker was an RB-36D-5CF, s/n 44-92088, of which only one model was built. On December 18, 1949, it made its first flight without the J47 jet pods, as was typical on the 'D' variant. But, following early testing, the jet pods were finally installed in time for President Truman's April 22, 1950, visit to Eglin AFB.

On June 3, 1950, the first RB-36D was handed over to the USAF. The RB-36D was assigned to the 28th Strategic Reconnaissance Wing at Rapid City AFB in South Dakota. From 1947 to 1950, the 28th SRW was entrusted with maintaining heavy bombardment proficiency and deploying tactical components as needed. From 1950 to 1955, its primary objective was global strategic reconnaissance, with bombing becoming a secondary duty. From 1955, the 28th SRW once again trained mainly as a bombardment wing but kept its reconnaissance capacity until September 1956. The wing operated the Convair B–36s 1949–1950 and the RB-36s 1950-1957.

Regarding the other RB=36Ds, after seven RB-36Ds were delivered, the final four aircraft on the contract's initial 100 aircraft, s/n 44-92095-44-92098, were constructed as B-36D bombers. On August 19, 1950, the first of them was deployed at Eglin AFB. The second production contract deliveries began from 49-2647 and included 22 B-36Ds and 24 RB-36Ds. The last, a B-36D-35-CF, was completed in June 1951.

Externally, the RB-36D model was no different from the B-36D. It did, however, have unique equipment to ensure optimal flying conditions for extended recon missions. Furthermore, the Air Force opted to convert the 22 B-36A types formerly employed for advanced training to RB-36Es. They also got 14 cameras, like the RB-36Ds, and four J47-GE-19 engines. Meanwhile, the second YB-36 prototype, which functioned as an experimental YB-36A for a period, was transformed into an RB-36E.

A close-up of the General Electric J47 axial-flow turbojet of a B-36. It was one of the most successful engines of the era, with a production total of over 30,000 units. Incidentally, it was also the first jet engine approved for commercial use in the United States. (DeGolyer Library, Southern Methodist University)

President Truman was a fan of air shows, and judging by his smile, he was a admirer of the B-36. The President is third from left, with glasses, and coat over his left forearm. This photo was taken at Andrews AFB, in Maryland. (Rowe. Abbie National Park Service. Harry S. Truman Library & Museum)

Convair RB-36Ds on the assembly line at the Convair plant in Fort Worth, Texas. (USAF)

A 'Featherweight II,' this B-36D, 44-92095, was delivered to the USAF in April, 1950. It was with the Air Research & Development Command's 6515th Maintenance Group at Edwards AFB and utilized for E-5 autopilot tests. On August 27, 1952, a ground fire at Edwards AFB damaged it. Then, on June 1, 1956, this Big Stick was scrapped at the Military Aircraft Storage and Disposition Center (MASDC). (T. Gianna, USAF)

This B-36 was sent to the USAF in August 1950. It was based at Carswell AFB in Fort Worth, Texas, together with the 436th Bombardment Squadron of the 7th Bombardment Wing. Sadly, on April 27, 1951, the B-36D was written off following a mid-air collision with an F-51 Mustang, 44-84973, while undergoing gunner training. The P-51 pilot and thirteen members of the B-36 crew perished. Four people managed to survive. (T. Gianna USAF)

Originally, this was a B-36A-15-CF, 44-92020, which had no armament, as it was exclusively used for crew training. In September 1948, it was delivered to the USAF. The bomber was eventually converted into an RB-36E-10-CF 'Featherweight II'. The 5th Strategic Reconnaissance Wing, Heavy, at Travis AFB, Fairfield, California, served as the RB-36's base of operations, from 1949 to 1950, and again in 1951. After that, it was transferred to the 72nd Bombardment Wing, at Ramey AFB in Puerto Rico, which had been activated on June 16, 1952, but, did not become operational until October 1952 . From March 1953 until March 1955, the 72nd's mission consisted of global strategic reconnaissance missions. In 1954, the mission had gradually shifted to one of bombardment training. The B-36 participated in those missions, together with other B-36s from the 72nd BW until 1956 . Finally, the Big Stick was scrapped on September 11, 1956 at MASDC. As for the other B-36s in the 72nd Bombardment Wing they were retired in 1958, and replaced by Boeing B-52s in 1959. (T. Gianna, USAF)

FICON

The Air Force's ten-engine giant Convair B-36D was the most potent bomber in the world in the early 1950s. It was the best deterrent to aggression, with an over-the-target speed of 435 mph and the ability to transport an atomic bomb across 10,000 miles. The RB-36D photo-reconnaissance derivative was similarly critical, as it was the only aircraft capable of taking key intelligence images of distant targets.

However, the B-36s were no longer safe in the skies, as the new jet in the Soviet Air Force, the MiG-15, proved to be a formidable fighter at the B-36's operating height of around 40,000 ft. Moreover, it was clear that strong defensive armament was no longer sufficient protection against fast-climbing jet interceptors or the first primitive anti-aircraft-guided weapons entering service during that era. Suddenly, bombers and reconnaissance aircraft had to fly significantly higher and quicker to avoid opposing fighters.

The USAF came up with an answer. The concept of carrying small or conventional fighter aircraft beneath a massive plane was not unique, as numerous countries had attempted the feat in the past. In fact, the first parasite aircraft flew in 1916 when the British launched a Bristol Scout fighter from a Felixstowe Porte Baby flying boat. However, the most significant programs in this field emerged in the United States in the mid-1940s.

Work on the McDonnell XP-85 Goblin, a small stout fighter, began shortly before the war's conclusion. It was a parasite jet fighter meant to keep the B-36 bombers safe. After the war, tests were undertaken, but the carrier was not a B-36 but a modified Boeing ETB-29A, 44-62093, that did the testing. Previously, the four-engine propeller bomber had taken part in a similar scheme with two Republic EF-84D Thunderjet fighters, s/n 48-641 and 48-661. The EF-84Ds were connected to the B-29's wingtips as part of the 'Tip Tow' project. That operation aimed to provide jet fighter protection for the bomber. However, the B-29's mobility with the added EF-84Ds on the wingtips proved to be only slightly advantageous, and the project was terminated.

The Air Force was sure that the parasite jet concept could be successful. Subsequently, in 1955, the program was revived under the new codename 'Tom-Tom.' This time, the carrier was a JRB-36F Peacemaker, 49-2707, and the jet was an RF-84F Thunderflash carried on the B-36's wing tips. That contributed to an increased wingspan-to-wing ratio, which boosted aerodynamic refinement and extended the range of the B-36. The 'Tom-Tom' project was widely tested in the 1950s, but it remained only experimental due to its overall complexity.

The aircraft in the left photo is a modified Boeing B-29 with two EF-84Ds attached to its wingtips. This was the 'Tip Tow' project, which was canceled due to the lack of maneuverability of the B-29. On the right, Convair JRB-36F Peacemaker, 49-2707, with an RF-84F, on its right wingtip. The left wingtip is devoid of any 'Tom Tom' jet. (USAF photos)

The USAF's 'Tip Tow' and 'Tom-Tom' experiments were disappointing. However, the Air Force felt compelled to keep one step ahead of any prospective enemy. As a result, it launched a third initiative known as FICON (Fighter Conveyor). This third project would have an RB-36F with a Republic RF-84F parasite fighter in its bomb bay instead of its wingtips. At least 12 RB-36Fs were modified by the USAF to 'GRB-36F standard, with a trapeze system in the bomb bay to carry, release, and collect the RF-84F.

The first tests were conducted in 1952. GRB-36F was the designation given to a production RB-36F-1-CF, 49-2707. That B-36 was the first to be modified with the specific trapeze system in its bomb compartment. In addition, a factory straight-wing F-84E Thunderjet, 49-2115, was outfitted with a retractable hook in the nose in front of the cockpit. The hook would connect the fighter to the trapeze, keeping the plane in the bomb bay during flight, lowering it for deployment, and hoisting it back when the mission was over.

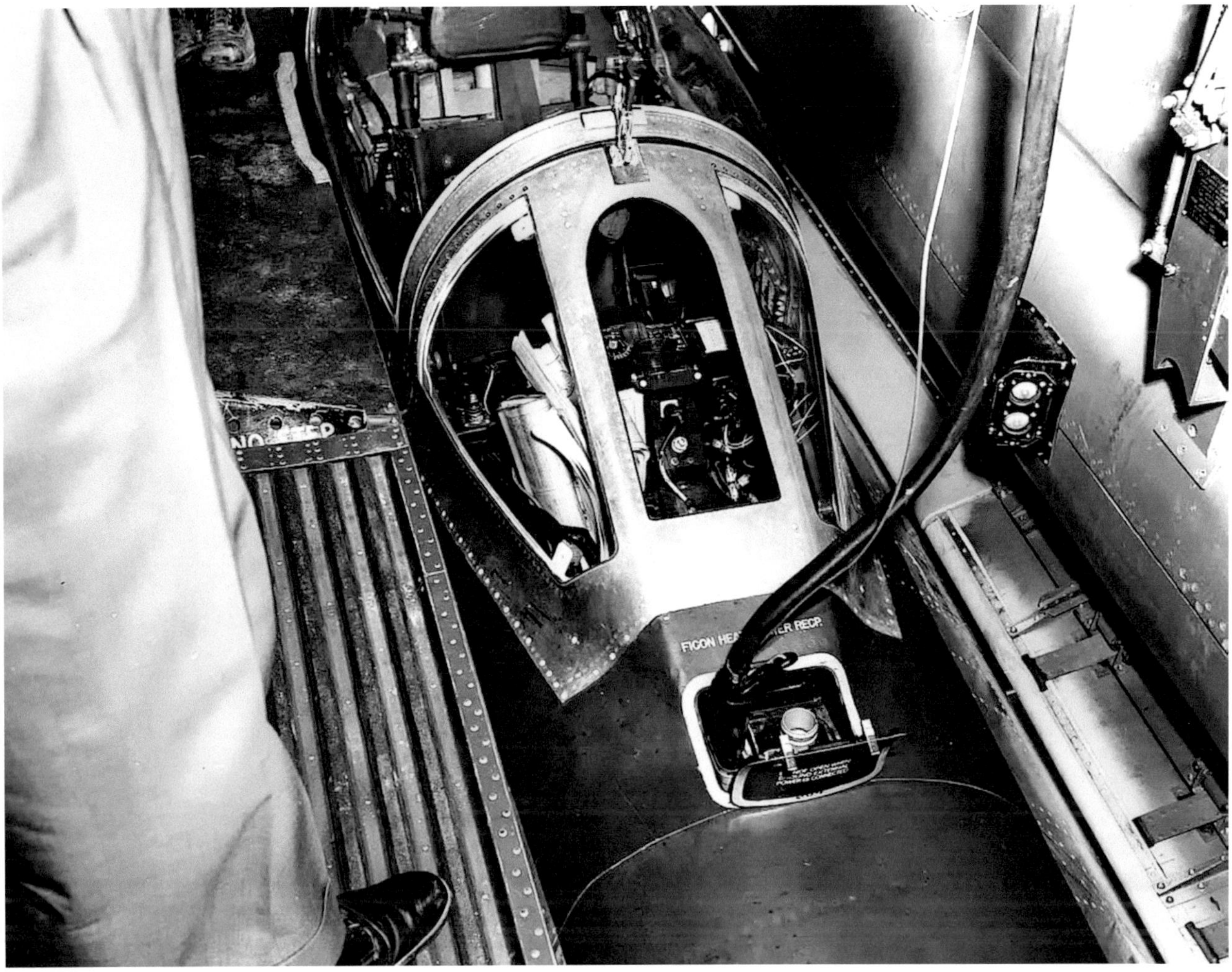

This is the canopy of a YRF-84F Thunderstreak, snuggly fitted in the bob-bay of the GRB-36F. (USAF)

Flight experiments began in May 1953 with the GRB-36F and the prototype swept-wing YRF-84F (initially designated YF-96A), s/n 49-2430. It was the first swept-wing jet fighter tested with the giant ten-engine 'Big Stick.' The crews from FICON, however, ran into a problem. The bomb bay of the B-36 could not easily accommodate the YRF-84F. However, by cutting away the bomb bay, the parasite jet could be cradled nicely in the GRB-36F, similar to how the Bell X-1 and Douglas Skyrocket research aircraft were carried during supersonic flight tests beneath a B-29 mothership. Apart from the B-36 alteration on the bomb bay, the prototype YRF-84F fighter's only significant modifications were to sweep its horizontal tailplane down in negative pitch and to place fittings on its nose and behind its cockpit to be snapped up by the 'trapeze' carried by the B-36.

The GRB-36F could take off with the YRF-84F in place, with the fighter pilot either in his cockpit or resting in the bomber. For the launch of the parasite aircraft, the fighter was lowered to the end of the trapeze. Its pilot would then start up the jet engine, and the plane would be uncoupled and dropped clear of the GRB-36F, like a bomb on its approach to a target. After completing its reconnaissance or bombing mission, the jet would reengage the trapeze boom with a probe on its nose. Then, another latch at the back of the boom would catch the hooks behind the fighter's cockpit, and the YRF-84F would be hauled up into the bomb bay of the B-36 once again.

The first GRB-36F, 49-2707, with an F-84-E 'seized' beneath the colossal bomber by its 'trapeze', like a praying mantis with a smaller insect. (T. Gianna, USAF)

The FICON program's tests were primarily successful. Both aircraft's equipment and systems were operational, and the GRB-36D's pilots efficiently managed the disengagement and re-engagement maneuvers of the RF-84F. As a result, in 1954, the USAF Strategic Air Command (SAC) put in an order for ten upgraded GRB-36Ds with a maximum speed of 440 mph (44-92090, 44-92092, 44-92094, 49-2687, 49-2692, 49-2694, 49-2695, 49-2696, 49-2701 and 49-2702).

In addition, twenty-five RF-84Fs were modified to GRF-84F standard and later redesignated RF-84K (s/n 52-7254 to 52-7278). The only differences between the RF-84K and the RF-84F were that the 'K' had a retractable hook on the nose in front of the canopy, its horizontal tailplane was in negative pitch, and some additional communications equipment.

The Strategic Air Command took possession of the FICON RF-84FK jets in 1955 and assigned them to the 91st Strategic Reconnaissance Squadron (91st SRS), based at Great Falls AFB, Montana (now known as Malmstrom AFB). The F-84Ks operated there for eighteen months until late 1957. The 91st Strategic Reconnaissance Squadron was deactivated in July 1957. It was time to retire the B-36s, as the new Lockheed U-2 reconnaissance aircraft were now available to the Air Force. With their excellent performance and simplicity, they quickly outstripped the parasite planes of the FICON program.

Interestingly, FICON aircraft were suspected of having flown some long-range penetration missions over the Eastern Bloc. However, there are no records to corroborate these claims. Finally, after the 91st SRS deactivation, the GRB-36Ds were retired, and the RF-84Ks were converted into standard reconnaissance aircraft without the 'hooks' in their noses, but the swept horizontal tail surfaces were retained.

This B-36 was delivered to the USAF in May 1951. It was initially an RB-36F-1-CF. However, in 1952 it was sent to the Convair plant at Fort Worth, Texas, for conversion to a GRB-36F-1-CF 'FICON' parasite carrier version and was fitted with the unique FICON trapeze. The Peacemaker's initial FICON trials were performed in 1952 with an F-84E, 49-2115. The first hook-up occurred on January 9, followed by the first retrieval into the bomb bay on April 23. Subsequently, its first attempt at the complete operation, from takeoff to landing, occurred on May 14 at Eglin AFB, Florida, in 1953. That was where the airborne launches and retrievals were usually performed. After the FICON operations, this B-36 was transferred to the 'Tom-Tom' project in 1955. Finally, it ended up at MASDC on May 9, 1957, where it was eventually scrapped. (T. Gianna, USAF)

This is another photo of the hardworking GRB-36F. It was taken the instant the RF-84F, 49-2430, became uncoupled from the enormous airframe above it. (T. Gianna, USAF)

B-36F

The following variant of the B-36 was the 'F' version. The B-36F model first appeared in late 1950 and was characterized by its six powerful Pratt & Whitney R-4360-53 engines, each developing 3,800 hp. This additional power significantly improved the performance of the B-36F compared to previous models and without excessive fuel consumption. Thirty-four B-36F bombers were fitted with the R-4360-53s; 49-2669-2675, 49-2677-2683, 49-2685, and 50-1064–1082. The only significant difference between the B-36F and the preceding B-36D was that the former used R-4360-53s while the latter used R-4360-41s. The prototype YB-36F, 49-2669, made its first flight on November 18, 1950.

The B-36F was also equipped with the K3A radar bombing system. Unfortunately, in the days before Bell Labs first produced transistors in New Jersey in 1947, bombing radars, like the K-3A, were operated with vacuum tubes, which were much larger and used significantly more power to run than a transistor. They also had a very short operational life. For example, in 1950, radios with vacuum tubes weighed at least 7 pounds and were usually bulkier than a loaf of bread or a shoe box. Even worse, they would need 7-pound batteries to work if they were portable.

That made the K3A radar bombing system a very clumsy one. For instance, on every B-36F flight, each aircrew member was burdened with carrying three-foot-long boxes filled with heavy replacement black boxes used as spares. They frequently used many of these spare pieces of equipment to keep the radar in operation to complete the mission. In addition, the B-36F was also equipped with the AN/APG-32 tail turret autotrack radar made by General Electric.

The Air Force received thirty-three B-36Fs in 1951; the final one was delivered in January 1952. In addition to the thirty-four B-36Fs ordered, the Air Force purchased twenty-four long-range RB-36F photo-reconnaissance models; 49-2703/2721 and 50-1098/1102. The RB-36F was distinguished by larger fuel tanks in its wings and a wide variety of photographic equipment. The recon model also had the K3A radar bombing system and the AN/APG-32 system.

The B-36F had a top speed of 417 mph and a service ceiling of 44,000 ft. with a standard combat load and 264,300 pounds combat weight. In contrast, the 24 RB-36F reconnaissance aircraft carried 22 crew members and 14 cameras. It had a top speed of 409 mph and a combat weight of 262,800 pounds.

The USAF received this Convair B-36F-1-CF, 49-2677, 'Featherweight II,' in August 1951. It was modified to serve as a carrier for the primary airframe and wing structure of the Convair XB-58. On March 12, 1957, the USAF flew the engineless XB-58 to Wright-Patterson AFB in Dayton, Ohio, for economic reasons. Installing engines and other subsystems on the XB-58 aircraft was estimated to be unnecessary during the tests. As a result, the cost-cutting measure was implemented. The XB-58 airframe with its protruding nose is clearly seen in the 'Big Stick's' belly. (T. Gianna, USAF museum photo)

Two Convair B-36Fs in flight. The nearest aircraft is B-36F-5-CF, 49-2683, 'Featherweight II' which the USAF received in August 1951. It was attached to 6th Bombardment Wing, Walker AFB, Roswell, New Mexico. (T. Gianna, USAF museum)

This is an underneath view of the XB-58 airframe being flown to Wright-Patterson AFB by the same B-36F that is on the opposite page, 49-2677. It is only using eight of its ten engines. The two inboard engines are lacking propellers as they were removed to avoid hitting the XB-58's wings below. (T. Gianna, USAF Museum photo)

This is another photo of B-36F-1-CF, 49-2677, carrying the XB-58 airframe in its belly. For obvious safety reasons, the flight was made with the inboard propellers removed, and with gear extended. (USAF - AFMC History & Heritage Program)

B-36F 'Featherweight III' Specifications

TYPE	POWERPLANT	DIMENSIONS WINGSPAN / LENGTH / HEIGHT	WEIGHTS Empty / Combat / Gross / Max. t.o.w.	SPEEDS Cruise / At 37,700 ft. Basic Mission / Combat / Max speed	SERVICE CEILING / RANGE
B-36F	Pratt & Whitney R-4360-53. 3800 hp. and General Electric J47-GE-19. 5200 lbs. thrust	230 ft/ 162 ft 0.25 in / 46 ft 10 in	167,646 lb. / 264,300 lb. / 370,000 lb. / 370,000 lb. (limited by structure).	235 mph / 421 mph / 413 mph / 421 mph Basic Mission	44,000 ft. / 7743 miles ferry range with 30,630 gallons of fuel
B-36F WEAPONS	2 x 20-mm M24A1 cannon, each in six retractable, remote-controlled fuselage turrets, a tail turret, and a nose mounting. Equipped with 9,200 rounds of ammunition. Normal bomb load: 72,000 lbs. Maximum bomb load: 86,000 lbs.				

THE GAS GUZZLER

Despite being the Air Force's pride and joy, the B-36 was not a very busy bomber. The Big Stick was the least active of the Air Force's three large bombers in 1951; the B-29, B-50, and B-36. The USAF's B-36 fleet flew a total of 23,499 hours in 1951, while the B-29s flew a massive 220,789 hours, and the B-50s completed 95,858,616 hours. That is a significant difference between the two Boeings and the Convair giant. The Peacemakers flew around the world 'making peace' wherever they went, while the poor B-29s and B-50s did the majority of the flying.

Furthermore, with only 23,599 hours flown, the B-36 consumed 21,556,733 gallons of fuel, compared to the B-29's 95,858,616 and the B-50's 36,855,308 gallons. Seemingly, the B-36 used less fuel, but it was in the air nine times less than the B-29 and just over three times less than the B-50. Therefore, the 'Big Stick' actually used much more gas. The rate of fuel per hour was 913 for the B-36, 434 for the B-29, and 507 for the B-50. Logically, with six propeller engines and four turbojets, the B-36 was indeed a gas guzzler.

Interestingly, the photo-reconnaissance variant, the RB-36, was less active than its B-36 sibling. In 1951, it only flew a measly 8,811 hours and consumed 8,327,458 gallons of fuel. However, its per-hour fuel consumption was 945, which was even higher than the B-36's.

Personnel from the B-36 test program pose on top of B-36F-5-CF, 49-2683, at Edwards AFB in 1955. This aircraft was delivered to the USAF in August 1951. A 'Featherweight II', it was with the 6th Bombardment Wing, Walker AFB, Roswell, New Mexico. It was scrapped on June 4, 1957, at MASDC. (USAF - AFMC History & Heritage Program)

B-36H

The B-36H design, which the Air Force evaluated based on an early joint manufacturer-user proposal, was a refinement of the B-36F model. The latest B-36H had a few modifications. It had a redesigned crew compartment, especially the flight engineer's station. Additionally, it was equipped with twin tail radomes to house the components of the new AN/APG-41A gun laying radar, which replaced the obsolete AN/APG-32 radar. There was also a change in the B-36-H's K system. It was relocated to the pressurized compartments to enable inflight maintenance.

Convair B-36H-1-CF, 50-1083, is getting a decontamination washdown by ground crew after air sampling tests during 'Operation Castle' at Bikini and Eniwetok atolls, in 1954. (National Archives)

The K-System was a Ballistics Control Panel that automatically measured the distance and time to the target. In addition, it determined the bomb's ballistics curve for the B-36's present altitude.

Externally, the main difference between the 'F' and 'H' models was the twin radomes of the AN/APG-41A tall gun system of the 'H' model.

Convair B-36H-1-CF, 50-1083, made its inaugural flight on April 5, 1952. The following month, it was delivered to the Air Force. It was based with the 4925th (Atomic) Test Group, Special Weapons Center, Kirtland AFB, in Albuquerque, New Mexico. This B-36 had an interesting time while it was with the 4925th.

The 4925th (Atomic) Test Group was founded at Kirtland AFB in New Mexico in February 1950. It was responsible for developing and testing the technology needed to provide the Air Force with nuclear weapons capability. It aided the Atomic Energy Commission and the Armed Forces Special Weapons Project in conducting nuclear weapons effects experiments. It also devised an operational technique for sampling those effects from the air. Also, it took part in the Buster-Jangle test series in November 1951. The Buster experiments were a weapons development operation, whereas the Jangle series assessed the utility of atomic weapons in cratering via ground-level and sub-surface explosions. The 4925th also participated in the Teapot test series of Operation Desert Rock at the Nevada Proving Ground, which lasted from February 18 to May 15, 1955. Operation 'TEAPOT' was an atmospheric nuclear weapons test series.

Seeing as an Air Force atomic testing group took in the first B-36H, the 4925th later used the 'Big Stick for air sampling tests during the 'Operation Castle' nuclear trials at Eniwetok and Bikini Atolls, in the Pacific Ocean, from February to May 1954.

After its atomic workout, the B-36H was transferred to the 7th Bombardment Wing at Carswell AFB in Fort Worth, Texas, and modified into a 'Featherweight III.' The bomber was eventually scrapped at MASDC on January 23, 1958.

The most perplexing aspect of the B-36H model was that it was the subject of a significant order. The Air Force was so taken with this version that it acquired a total of 156 aircraft, 83 B-36Hs, and 73 RB-36Hs. That accounted for almost 40% of all B-36s sold. The enormous order astonished numerous people. The 'H' variant was not significantly different from the 'F' version, of which only 58 were sold. Yet, it was the most successful of all B-36 types. The B-36Hs were delivered between December 1951 and September 1953.

In March, only three months after the deliveries of the B-36H began, defective landing gears caused a series of accidents. As a result, except for the initial 152 B-36s, the Air Force grounded all B-36s. That practically meant all of the B/RB-36Fs manufactured in the second batch and the recently delivered 30 B/RB-36Hs the Air Force had already accepted would remain on the ground. From the outset, investigators blamed the aircraft's landing gear pivot shaft for the accidents. As a result, the maintenance personnel created a heavier bar as a stop-gap measure until a permanent replacement could be made. As a result of this measure, the grounding orders were lifted.

In early 1952 the Strategic Air Command needed tankers, B-36H tankers to be precise. The USAF sought the B-36Hs as tankers for high-altitude refueling of jet fighters since the Peacemakers could fly at high altitudes and great speeds for their size compared to other tanker aircraft at the time. Keep in mind that those were the days before the magnificent KC-135s arrived on the scene.

The USAF then requested that Convair install a probe and drogue refueling system on one B-36. The contract for the alteration was granted in February 1952, and work was finished in May. However, testing was later put off due to the late arrival of the B-47 receiver plane. Finally, after an improved British-made probe and drogue refueling system was installed, tests were resumed in January 1953. The British system allowed the modified B-36H tanker to refuel many receiver aircraft simultaneously. Moreover, within 12 hours, if needed, the 9-crew tanker could be converted back to a standard bomber configuration.

The proposal to use numerous B-36H refueling aircraft as tankers could not be sustained over time. The B-36's primary mission was to remain dedicated to its function as a bomber rather than providing fuel to thirsty jet fighters. In light of this, the B-36H tanker concept never took off.

Finally, the end of production for the B-36Hs was in July 1953. The final B-36H was handed over to the Air Force at the end of September of that year.

A 43,000 pound bomb on display in front of a B-36H during the Armed Forces Day open house at Carswell AFB on May 20, 1950. (USAF - AFMC History & Heritage Program)

(National Archives)

Convair B-36H-1-CF, 50-1085 (photo above) seen taxiing at Eglin AFB in the 1950s. This bomber was used in support of 'Operation IVY,' a series of two atmospheric nuclear tests conducted by the Atomic Energy Commission (AEC) at Enewetak Atoll in the Marshall Islands from November 1 to December 31, 1952. The series included two detonations. The first, dubbed 'MIKE,' was carried out on November 1, and the second, nicknamed 'KING,' was detonated on November 16. Previously, in 1950, the US government chose to explore the development of thermonuclear weapons, and the 'IVY' series was a critical step in this process. 'MIKE' was an experimental device that created the first thermonuclear detonation by generating a significant portion of its energy by the fusion, or joining, of hydrogen atoms. 'KING' was a stockpile weapon modified to create a high yield. The B-36H bomber released it. The energy from 'KING' was generated by the fission, or splitting, of plutonium atoms. At the time, these were the greatest-yield nuclear detonations.

Later, this B-36H was involved in another project. It was converted to EDB-36H-1-CF to test the GAM-63 'Rascal' missile. The Rascal, originally known as the 'XB-63,' was a nuclear-armed air-to-surface supersonic guided missile. The missile was designed to be a 'standoff' weapon, to be fired from Strategic Air Command (SAC) bombers up to 100 miles away, limiting the manned bomber crew's exposure to enemy defenses in the immediate target area. Even though the tests were more than satisfactory, the program was terminated in late 1958, shortly before the first Rascal-equipped SAC unit was to become operational. The reason was that the USAF favored the more promising and more extended range 'Hound Dog' missile.

A B-36H climbs above the clouds, over the countryside, somewhere in the US. (T. Gianna, USAF)

Crew in a B-36H-5-CF, 50-1110, 'Featherweight III', ensconced in the lower flightdeck, inside the nose area of the bomber. (USAF - AFMC History & Heritage Program)

This is a B-36H-35-CF variant, 51-5730, which was one of six produced, and was a 'Featherweight III'. It is flying over the Atlantic ocean with an unusual cloud formation, on October 28, 1956. It was delivered to the USAF in October 1952, and joined the 7th Bombardment Wing, Heavy, based at Carswell AFB, Fort Worth, Texas. The 7th BW,H began converting to B-36 bombers in June 1948. The Wing Controlled two B-36 groups, and operated three B-36 squadrons, from February 1951-May1958, when the Big Sticks were decommissioned. The Wing also evaluated an RB-36 during 1949/1950. (USAF - AFMC History & Heritage Program)

RB-36H-40-CF, 51-13739, 'Featherweight III', on exhibit at San Francisco in August 1954. A rare sight with the gun turrets up. This B-36H was delivered to the USAF in December 1952, and was with 5th Strategic Reconnaissance Wing, Heavy, when this photo was taken. The 5th SRW,H operated RB-36s from 1951–1958. It also operated B-36s from 1955-1958. The Wing was redesignated 5th Bombardment Wing, Heavy, on October 1, 1955. All the B-36s were replaced by B-52s in 1959. (Bill Larkins)

Convair RB-36H soaring like an albatross across a dark sky with its massive wings. The dual radomes on the tail turret were a feature of the enhanced AN/APG-41A fire control radar system found on the 'H' and 'J' models. (USAF Museum)

This B-36H-1-CF, 50-1087, was the fifth 'H' model built. It was delivered to the USAF in May1952. A 'Featherweight II', it was with the 95th Bombardment Wing, at Biggs AFB, El Paso, Texas. Scrapped in 1958. (USAF Museum)

THE ATOMIC XB-36H/NB-36H

The USAF announced requirements for a new system known as the 'WS-125 A Weapon System' in 1954. It sounded like any other standard Air Force requirement, but it wasn't. The 'WS-125 A Weapon System' was a United States Air Force project to develop a super long-range strategic bomber that would be nuclear-powered or have combined-energy power. In any case, it was supposed to have unlimited endurance; anything less would not be good enough.

In 1956, the two largest engine manufacturers and two biggest airplane makers competed for the project. General Electric joined forces with Convair on one side, while Pratt & Whitney teamed up with Lockheed on the other. The two competitors got right to work on the new atomic-powered bomber. The engine manufacturers were in charge of the powerplants, while the two aircraft manufacturers would take care of the airframes.

General Electric was the first to announce its 'X211' engine, whose military designation was J87. It was made up of two jet engines and a single reactor. Because the air heating in the power plants was indirect, it could be used without fear of radioactive contamination. Two shielded reactors successfully powered two General Electric J87 turbofan engines to nearly full thrust.

Pratt & Whitney first proposed a blower engine, with a 3-meter diameter blower driven by a steam turbine in a closed system, with hot air fed from the cooling reactor behind the blower. Pratt & Whitney then changed its mind and decided on a second atomic engine identical to General Electric's J87. The P&W engine was designated J91, which comprised six units with turbines and compressors and one main closed-cycle reactor.

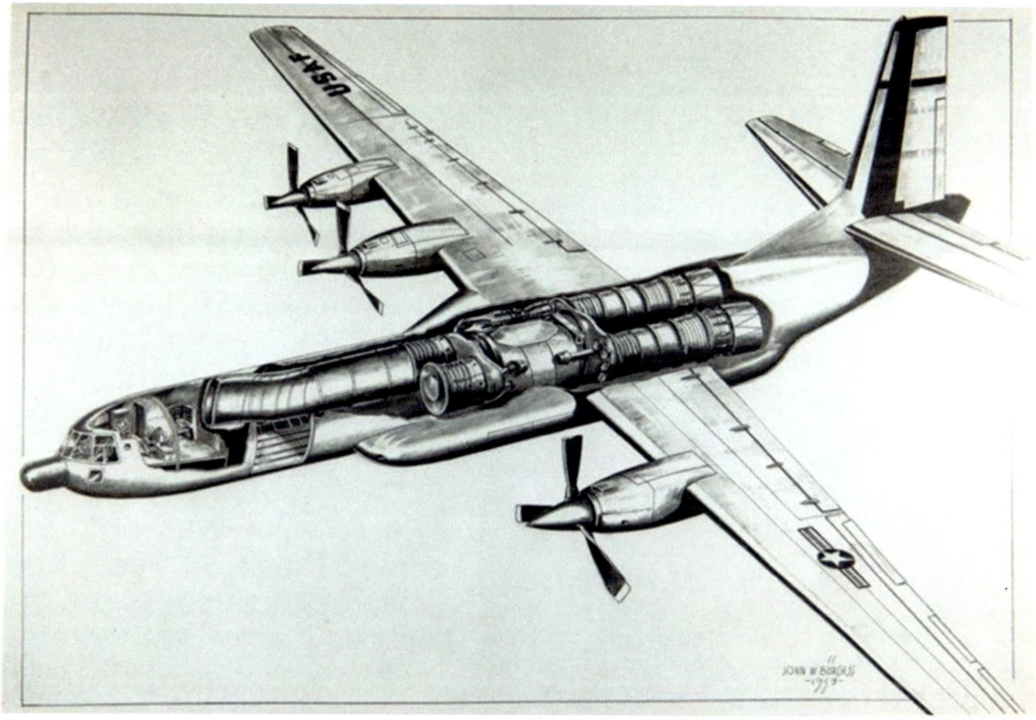

This is a cutaway drawing of the proposed WS-125 testbed aircraft, showing the General Electric J87 nuclear powered engines. The aircraft looks eerily like a three-engine Douglas C-133. (John W Burgess/Public Domain)

While the two engine makers, and two plane manufacturers, were doing tests, the USAF had been doing its own experiments with nuclear power. In 1954, when it first announced the 'WS-125 A Weapon System' project, the Air Force found a testbed aircraft for its project in the form of a B-36H-20-CF, 51-5712.

The B-36 bomber was delivered to the USAF in June 1952 and was assigned to Carswell Air Force Base in Fort Worth, Texas. Unfortunately, less than three months after its arrival, Caswell AFB suffered probably the worst time in its history. On September 1, 1952, Labor Day, Caswell was utterly devastated by a powerful tornado that cut through the base like a hot knife through butter. One B-36 was totally destroyed, and 82 others were heavily damaged. Luckily, the Air Force and Convair quickly assessed the damage, and repairs began in earnest. Within a week, the first B36 returned to duty, and nine more the following week. The last damaged aircraft returned to operational duty in May 1953.

Some of the damage at Caswell AFB caused by the Tornado in September 1952. (USAF - AFMC History & Heritage Program)

The Air Force discovered their testbed, not in a lovely clean, expansive Air Force Base, but among the rubble of Caswell. That day, one of the heavily damaged Peacemakers was Convair B-36H-20-CF, 51-5712. The forward part of its fuselage looked like Thor's Hammer came down on it a dozen times. Luckily the Air Force extracted the B-36 from the disaster of Caswell, and after a lengthy repair process, the bomber was ready to enter service again.

In early 1953, the Air Force assigned the B-36 to Project MX-1589 (the fuse of a nuclear reactor on board a modified aircraft), and it was nominated to be the project's Nuclear Test Aircraft (NTA). The Convair XB-36H was the world's first, and most likely the last, aircraft to fly with an operational atomic reactor. It was modified to carry a 3-megawatt, air-cooled nuclear reactor in its bomb bay. The reactor was not designed to power the B-36, like in the 'WS-125 A Weapon System' project aircraft. Instead, a low-powered Convair-built nuclear reactor was installed on the plane for conducting tests. The main objectives of the airborne experiments were to confirm ground reactor data, gain experience in reactor handling and equipment operation, research crew shielding, and the overall effects of radiation on the airframe. Regarding the shielding of the crew, they were placed in a lead-encased cockpit, which looked like a giant bee hive or cocoon. However, it did the job.

Air Research and Development Command designated the test plane XB-36H-20-CF on March 11, 1955. Its first flight for the Command took place on July 20, 1955, and it was labeled NB-36H-20-CF on June 6, 1956.

The nuclear NB-36H had a redesigned fuselage nose and flight deck. It was also equipped with air scoops on each side of the rear fuselage, just aft of the wings, which supplied cooling air to the reactor when it was in operation. The reactor was turned on only when the NB-36H was high over unpopulated areas. All flights were conducted over remote regions of the Southwest United States, mainly over New Mexico and Texas, to avoid endangering the general population.

First flown on September 17, 1955, the NB-36H had a dark blue nose and an orange radiation symbol on its tail, distinguishing it from all other B-36s in service. It was initially given the name 'Convair Crusader' painted on the blue fuselage flashes, but Convair dropped that name to avoid confusion with the US Navy's supersonic Chance Vought F8U-1 'Crusader'.

The NB-36H was accompanied on every test flight by a Boeing B-50 escort. That was because the technical personnel required for the tests were flying in the B-50 formation at a safe distance. The B-50 crew also included ten engineers who had been trained as parachutists. They were members of the safety team in charge of these uncommon tests. In an accident involving the NB-36H, it would be their responsibility to jump near the accident site and secure it.

The USAF eventually canceled the Nuclear Aircraft Program in 1958, and the NB-36H was scrapped in Fort Worth, Texas, the same year. Throughout the mission, the NB-36H completed 47 flights and 215 hours of flight time, with the reactor operational for 89 of those hours. The entire campaign ran from September 17, 1955, to March 1957.

B-36H Specifications

TYPE	POWERPLANT	DIMENSIONS WINGSPAN / LENGTH / HEIGHT	WEIGHTS Empty / Combat / Gross / Max. t.o.w.	SPEEDS Cruise / At 34,500 ft., Basic Mission / Combat / Max speed	SERVICE CEILING / RANGE
B-36H	Pratt & Whitney R-4360-53, 3800 hp, and J47-GE-19, 5200 lbs. thrust	230 ft/ 162 ft 0.25 in / 46 ft 10 in	168,487 lb. / 253,900 lb. / 370,000 lb. / 357,500 lb. Limited by strength	234 mph / 402 mph / 413 mph / 416 mph	44,000 ft. / 10,000 lbs. bombs and 1,408 lb. chaff, 7,691 miles ferry range
B-36H WEAPONS		2 x 20-mm M24A1 cannon each in six retractable remotely-controlled fuselage turrets, tail turret, and nose mounting, with 9200 rounds of ammunition. Normal bomb load up to 72,000 pounds. Maximum bomb load 86,000 pounds.			

The Caswell tornado survivor, NB-36H, with its new nose, took its first flight on September 17, 1955. (USAF - AFMC History & Heritage Program)

This is the bee-hive, or cocoon like, crew capsule used on the NB-36H, which could accommodate a crew of five. It was lead-lined, with specially designed 18-inch thick windows in the front windscreen. It weighed 11 tons. Despite its unusual design, it kept the plane's crew safe and secure. (USAF)

This photo shows the NB-36 banking slightly to the left, over a sparsely populated area. (T. Gianna, USAF Museum)

A close up look of the nose section of the NB-36H. Note the aircraft has its original designation; XB-36H. (USAF Museum)

The Convair NB-36H experimental aircraft with the nuclear reactor, 51-5712, looking majestic flying above the clouds with the Boeing B-50 chase plane keeping a slight distance away. The photo was taken during research and development tests at the Convair plant in Fort Worth, Texas. Note the nickname of the NB-36H 'Convair Crusader'. This designation was dropped by Convair to avoid confusion with the US Navy's supersonic jet fighter F8U-1 'Crusader' which made its first flight in 1955. The same year this photo was taken. (USAF)

B-36J

The final B-36 variant was the B-36J. The first aircraft, 52-2210, took to the skies on September 3, 1953. The B-36J version had additional fuel tanks in the outer wings, allowing the bomber to carry an extra 2,770 gallons of fuel. That had long been the goal of SAC. The USAF and Convair considered boosting the takeoff weight of available B-36s without affecting the safety of the crews as early as 1952. However, according to USAF engineers, the structural stability of several of the aircraft's new components remained uncertain. Then, when the 'J' version appeared, the Air Force had the undercarriage strengthened to accommodate the aircraft's new gross weight of 410.000 lb., making it heavier than any other aircraft in the inventory.

Deliveries of 33 aircraft, 52-2210/2226 and 52-2812/2827, began in October 1953. On Tuesday, August 14, 1954, SAC personnel took delivery of B-36J number 52-2827, the last of the 33 aircraft in the 'J' series and thus the 383rd and final B-36 produced.

In 1954, 'Project Featherweight' was a scheme that aimed to lower the B-36's empty weight to increase its speed, altitude, and range. Subsequently, Class II Featherweights retained all of their weaponry. However, the last 14 B-36Js entered service as lightweight B-36J-IIIs. Class III aircraft removed all their gun turrets and fire control systems, save for the tail guns. Moreover, they had 13 crewmen. The first delivery of the B-36J-IIIs was in February 1954. The 'J III' had a combat range of about 4,000 miles, a top speed of 418 mph at 37,500 ft., and a service ceiling of 43,600 ft. The weight reduction was palpable. Empty weight fell from 171.035 lb. to 166,165 lb.; however, maximum take-off weight stayed unchanged at 410,000 lb.

This is Convair B-36J-75-CF, 52-2827. Nine B-36s of this model were built. This aircraft was the last B-36 built and was handed over to the USAF on August 14, 1954. It was assigned to the '92nd Bombardment Wing', Fairchild AFB, Spokane, Washington. This Wing was previously known as the '92d Bombardment Wing, Medium'. Then, on June 16, 1951, the Wing was redesignated as the '92d Bombardment Wing, Heavy'. The Wing operated B-36s from 1951 to 1957, and the B-52s subsequently replaced them. The last wing this B-36J was assigned to was the '95th Bombardment Wing, Medium' at Biggs AFB, Texas. That Wing was established on June 4, 1952, and was activated on June 16, 1952. Five months later, on November 8, 1952, it was redesignated as the '95th Bombardment Wing, Heavy'. It operated B-36s from 1953–1959. That was where this last B-36J spent its final days of service. (Mike LaChance)

Between February 11 and December 31, 1954, several earlier B-36 models also underwent featherweight modifications. The B-36F (III) and B-36H (III) were attributed with the most incredible B-36 performance, with 423 mph top speed and 47,000 ft. service ceiling. However, the B-36D (III) Featherweight could also attain 418 mph in this configuration. Finally, August 14, 1954, saw the delivery of the last B-36J (III), marking the end of manufacturing at the Fort Worth plant in Texas. In the end, 383 B-36s were delivered to the Air Force, including the two prototypes.

Even though new aircraft deliveries ceased in August 1954, the Air Force's SAM-SAC program (Specialized Aircraft Maintenance-Strategic Air Command) modified each B-36 to the most up-to-date configuration. This procedure was carried out till April 1957. Due to the conversion of all RB-36s from reconnaissance to bomber aircraft, all four Strategic Reconnaissance Wings were renamed Heavy Bomb Wings on October 1, 1955. At the height of the B-36's power in 1955, the Strategic Air Command had ten Wings in operation, consisting of 209 B-36s and 133 RB-36s.

Even so, by the mid-1950s, the 'Big Sticks' obsolescence was all too apparent. The B-36s time was up. In 1955, the B-52s made their grand entrance and immediately started replacing the B-36s. SAC's final B-36, and the last produced, 52-2827, departed the 95th Bombardment Wing on February 12, 1959. The last B-36 to take to the skies was B-36J-65, 52-2220, which was flown to the Air Force Museum in Dayton, Ohio, in April 1959. The B-36s never made it to the decade of the turbulent 1960s. Thirty-three B-36Js were made, and the total of all B-36s produced was 383 units.

This photo depicts the festivities for the final B-36 to leave the production line, on August 14, 1954. The event drew approximately 11,000 people to Convair's manufacturing plant in Fort Worth, Texas. This Peacemaker is one of just four remaining, and intact B-36 airframes. It is presently on display at the Pima Air & Space Museum in Tucson, Arizona. (USAF - AFMC History & Heritage Program)

A photo of the B-36J's cockpit, taken at the National Museum of the Air Force. (USAF Museum, Ken LaRock)

Convair B-36J-75-CF, 52-2827, taking off from an undisclosed Air Force Base. (USAF)

The photo above is of Convair B-36J-1-CF, 52-2220, originally a B-36J-65-CF. The 'J' models had the same contract numbers and serial numbers as the 'H' variant; therefore, they were listed at Convair with new numbers, specifically 'B-36J-1-CF'. This particular aircraft was delivered to the USAF in January 1954. It was stationed at Carswell Air Force Base in Fort Worth, Texas, as part of the 98th Bombardment Squadron, Heavy. The Squadron was previously known as the '98th Bombardment Squadron, Very Heavy' before being redesignated on December 1, 1948, to its current title. From 1949 through 1957, the 98th BS/H flew B-36 bombers.
B-36J, 52-2220, was involved in a severe accident when it was assigned to the 98th Bombardment Squadron, Heavy. On April 30, 1954, it suffered heavy damage in extreme weather conditions.
Then, in December 1954, the B-36J was reassigned to the 42nd Bombardment Wing, Heavy, at Loring AFB in Maine. The 42nd BW, H, was established on February 19, 1953, and was activated six days later, on February 25, 1953. The 42nd operated B-36s from 1953 until 1956. After the B-52s replaced the B-36s in 1957, the B-36J was relocated again. This time it was to the 5th Bombardment Wing at Biggs AFB in El Paso, Texas, where it was modified to a 'Featherweight III.'
On February 3, 1959, the B-36J was sent to MASDC, although it wasn't scrapped. It was one of the lucky B-36s to avoid the executioner. The bomber was sent to the National Museum of the US Air Force in Dayton, Ohio, where it still resides. This photo shows the 'Big Stick' being moved through the Museum grounds, from building one to building three, in October 2002. (USAF Museum)

B-36J Specifications

TYPE	POWERPLANT	DIMENSIONS WINGSPAN / LENGTH / HEIGHT	WEIGHTS Empty / Combat / Gross / Max. t.o.w.	SPEEDS Cruise / At 36,400 ft., Basic Mission / Combat Basic Mission / Max speed	SERVICE CEILING / RANGE
B-36J	Pratt & Whitney R-4360-53 radial of 3,800 hp (takeoff power), General Electric J47-GE-19 turbojet of 5,200 lb. thrust	230 ft/ 162 ft 0.25 in / 46 ft 10 in	171,035 lb. / 266,100 lb. (Basic Mission) / 410,000 lb. / 410,000 lb. Limited by strength (Landing gear and wings)	227 mph /410 mph / 410 mph / 435 mph	44,000 ft. / 7,144 miles ferry range
B-36J WEAPONS		6 M24 20mm cannon in 8 nose, tail, and fuselage turrets. Bombs: Either nuclear or conventional bombs. Normal bomb load: 72,000 lbs. Maximum bomb load: 86,000 lbs.			

CONVAIR XC-99

In September 1942, the Army Air Force signed a contract with Consolidated-Vultee for the production of a strategic transport/cargo version of the, yet to be made XB-36. The new aerial 'heavy-hauler' version of the B-36 bomber would have the same engines, wings, tail assembly, and landing gear as the Peacemaker. The initial study of the cargo version was given the code name of 'Model 37', and a prototype was ordered with the designation 'XC-99', 43-52436.

In 1943, a full-size wooden mock-up model of the Convair XC-99 was built. However, due to lengthy delays and interruptions caused by more urgent programs, the prototype did not make its first flight till November 23, 1947, when it took off from Lindbergh Field in San Diego, California. The pilot at the controls was Russell R. Rogers, and the co-pilot was the well-known Beryl A. Erikson.

Once it appeared in front of the media and the general public, they were genuinely fascinated by XC-99's size. They saw a plane with a vertical stabilizer 57 ft. 6 inches high, about two ft. shorter than the Presidents' heads on Mt. Rushmore. Furthermore, its fuselage (182 ft. 6 inches) was almost as long as the length of two Antarctic blue whales placed end to end, and with a gross weight of 265,000 lb., the XC-99 was just 11,500 lb. lighter than the B-36A bomber at 276,500 lb.

The XC-99 had a fuel capacity of 21,000 gallons. Its maximum takeoff weight was 322,000 pounds, distributed among ten tires, allowing the massive aircraft to land (or take off) on any 5,000-foot runway that could sustain the weight of the considerably smaller C-54 transport/cargo aircraft.

The XC-99 was made to carry hefty weights, which it did with ease, as its voluminous interior provided enough space. Therefore, with its normal gross weight of 265,000 lb., the Convair XC-99 could carry a military load of 100,000 lb. over 1,615 miles or 44,000 lb. over 4,720 miles. In maximum takeoff weight mode, at 320,000 lb., the six-engine beast was able to carry 100,000 lb. of cargo over 2,980 miles., 44,000 lb. over 6,210 miles, or 11,000 lb. over 8,075 miles.

The XC-99's interior was 180 ft. long, 26 ft. wide, and 14 ft. high, with two decks. That allowed the massive transport to carry 400 fully equipped troops or 300 wounded on stretchers. The enormous cabin of the XC-99 was also suitable for hauling heavy equipment. Additionally, various types of vehicles could be easily placed on the lower deck, which had direct access via ramps incorporated into the aircraft's hatches.

The Convair XC-99 had a cantilever wing (a wing that uses no external struts or bracing), just like the B-36. However, the B-36A and the XC-99 also had an aspect ratio of 11 (see below). That was a considerable aspect ratio for the XC-99 due to its size, with Its total wing area measuring 4,768 square ft. In addition, the wing of the B-36 bomber was semi-high and rested on the back of the fuselage. On the other hand, the wing of the XC-99 was median, meaning it was embedded at the mid-height of the aircraft's fuselage. Apart from that, the two wings had the same dimensions.

(High aspect ratio= long narrow wings give the plane less induced drag, more stability, and fuel efficiency compared to Low aspect ratio= shorter wings, the aircraft uses more fuel but is far more maneuverable)

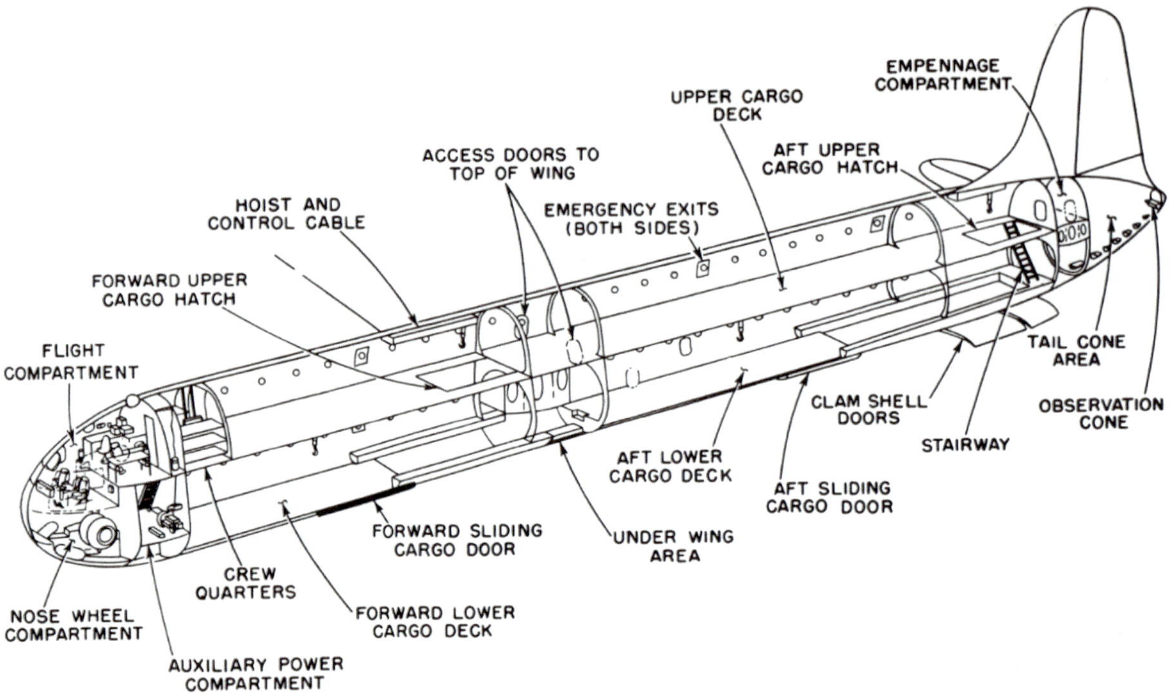

Interior details of the single USAF Convair XC-99, 43-52436. (USAF Museum)

Convair XC-99 at San Diego, California, circa 1948. (National Archives)

XC-99 IN OPERATION

After the inaugural flight in November 1947, Consolidated Vultee tested the XC-99 in its San Diego and Fort Worth facilities until May 26, 1949, when it was accepted as an experimental aircraft by the USAF and assigned to the Eighth Air Force (SAC) in Fort Worth, Texas. Following approval, the plane was grounded for a year while structural changes were made to increase the XC-99's permissible gross weight. The XC-99 stayed with the Eighth Air Force at Fort Worth until September 8, 1950, when it was moved to Kelly AFB, which served as the primary B-36 supply base. The testing program resumed once the aircraft arrived at Kelly AFB.

In June 1949, the XC-99 created a record when it took off from the Fort Worth Division airfield with cargo weighing 100,000 lb. It was thought to be the heaviest load ever carried by a single plane.

The Postal Service was highly interested in the Convair XC-99's potential employment as an aerial 'postal station' during the aircraft's January 1951 test flights. Convair predicted that the XC-99 could be operated commercially by the Postal Service between New York and Chicago and between Chicago and New Orleans or Los Angeles, based on its cargo capacity of 16,117 cubic ft. and its ability to carry 100,000 lb. During the Postal Authorities and Convair discussions, the XC-99 was undergoing trials with the Air Force as a military freighter. It transported over one million pounds of cargo in its first seven weeks of operation. These missions included a transcontinental flight with a payload of 85,000 lb. The Post Office considered leasing the XC-99 from the US Air Force to conduct tests with heavy mail loads between major distribution hubs in the United States. However, nothing materialized despite the Postal Authorities' eagerness to continue with the concept.

The Postal Service's disappointment was, in all probability, due to the Korean War. During that conflict, the XC-99 began conducting twice-weekly flights from Kelly AFB to McClellan AFB in California. McClellan AFB served as a vital supply base during the Korean War, and the XC-99 provided McClellan AFB with its needs. But, war was serious business; the mail could wait.

A big battle went on between the XC-99 and its Douglas rival, the C-124 Globemaster II, to see which of the two titans could transport the heaviest weights. Subsequently, in June 1951, both aircraft created new records. The enormous XC-99 carried 85,000 lb. of cargo on a 2,600-mile non-stop flight from Sacramento, California, to Savannah, Georgia, and achieved a new distance mark at that load. The smaller C-124 Globemaster II hoisted a 70,000 lb. cargo, flew 1,000 miles, unloaded it, and returned without stopping for fuel. This load increased the plane's takeoff weight to 210,000 lb., 17 tons over its designed gross weight.

Thanks to the heavy 'logistics' cargo planes like the C-124 Globemaster II, and the solitary Convair XC-99, with its six Pratt and Whitney Wasp Majors, American planners foresaw strategic transports capable of carrying 100,000 lb. payloads over transatlantic distances in the future. The XC-99, which was earlier regarded as a white elephant, proved itself a most valuable asset for the Air Force. By August 1954, it had logged 4,105 flying hours on 508 flights, carried over 36 million lb. of cargo, and reduced operating costs to 11.2 cents per ton-mile. On two particular flights, it took off fully loaded from a 5,400 ft runway on the first flight and outdid itself on the next one, when it took off with a payload of 104,000 lb. from a runway at an altitude of 5,000 ft. On another trip, it hauled the complete ground equipment of a fighter wing

2,330 miles non-stop across the American continent! If that did not prove the power of the beast, nothing would.

In other record-breaking flights, the XC-99 successfully completed an airlift from Dover, Delaware, to Iceland. It flew 200 hours during the month and required only the changing of one engine. As a result, original cargo loads set at 300,000 lb. were exceeded by 79,000 lb.

Although the XC-99 was always regarded as an experimental aircraft, the experimental 'X' designation was unofficially removed in August 1953 on a flight from San Antonio to Rhein-Main, West Germany. That was due to Government regulations restricting experimental aircraft from traveling overseas. However, by July 1954, the XC-99 was no longer considered a project aircraft and was maintained and scheduled as a regular aircraft.

From 1955 onward, the solitary XC-99 Goliath was losing its importance in the eyes of the Air Force. As a one-of-a-kind aircraft, the XC-99 was utterly dependent on its sister design, the B-36, for the supply of spare parts. However, as the USAF's B-36 inventory started to dwindle in the mid-1950s and new B-52s began to take their place, maintaining the XC-99 grew more challenging. Furthermore, USAF officials believed that the initial XC-99 program was no longer justified in the context of the time. The USAF possessed a fleet of good military transport planes in the mid-1950s, which were undoubtedly more modest than the XC-99, but they successfully met the Air Force's objectives. It should also be noted that at $4.7 million, the XC-99 was far too pricey. As a result, the Air Force put the XC-99 to the sword. All XC-99 flights were canceled in March 1957. It was predicted that 144,734 hours of labor and more than $1 million would be necessary to replace the structural components of the XC-99 that were no longer repairable. Because of the high operational costs, the USAF declared the XC- 99 surplus on July 11, 1957, after three years and 7,400 hours of service. The Convair XC-99 would have been the largest prop-driven cargo transport aircraft in Air Force history if it had been put in production, but the Douglas C-133 Cargomaster holds that honor.

XC-99 Specifications

TYPE	POWERPLANT	DIMENSIONS WINGSPAN / LENGTH / HEIGHT	WEIGHTS Empty / Combat / Gross / Max. t.o.w.	SPEEDS Cruise / Average Speed for Range / Max speed	SERVICE CEILING / RANGE
XC-99	Pratt & Whitney R-4360-41 Wasp Major 28-cylinder air-cooled radial piston engine. 3,500 hp	230 ft / 182 ft 0.25 in / 57 ft 6 in	135,232 lb. / 162,250 lb. / 320,000 lb. / 300,000 lb.	185 mph Basic Mission / 188 mph Ferry Mission / 307 mph	40,490 ft. / N/A
XC-99 WEAPONS		N/A			

The sole prototype XC-99 off La Jolla Cove, in San Diego, California. Seen here in its early days of operation, before the fitting of a nose radome, and four-wheel main gear bogies. (T. Gianna, USAF Museum)

The XC-99 in 1953. Note the AN/APS-42 weather radar on the front with the distinctive black 'thimble like' radome. (T. Gianna, USAF)

Here's a tale of two fates. The XC-99 was Convair's unique model that didn't make the cut and faded into obscurity in 1957. On the other hand, the small Convair liner, and its sibling Convair prop liners were produced and sold in the hundreds to airlines and military forces around the world. (USAF - AFMC History & Heritage Program)

This photo displays the airline configuration of XC-99's upper deck. After the USAF wrote off the XC-99, Convair was disappointed and sought to present the 'Model 37' to many US airlines as a passenger transport option for long-distance high-traffic routes, but only Pan Am showed interest. The airline initially ordered 15 of the 'Super Clippers' but canceled the order when the postwar economy hit a slump. With its spacious interior, the Model 37 would have easily carried over 200 passengers. The jet-powered 707, which first flew on December 20, 1957, couldn't carry that many travelers. (USAF - AFMC History & Heritage Program)

CONVAIR YB-60

On August 25, 1950, Convair issued a formal proposal for a swept-winged version of the B-36 with all jet propulsion. The USAF was sufficiently interested that on March 15, 1951, the Air Force authorized Convair to convert two B-36Fs, s/n 49-2676 and 49-2684, as B-36Gs. The new model was dissimilar to the pre-existing B-36; due to Convair's numerous alterations, it was designated YB-60 in the middle of 1951. Subsequently, Convair's plans to bring existing B-36s to the 'G' configuration were all but over. There was to be no 'G' version.

In August 1951, Convair proposed building the two prototype aircraft with alternative configurations because the final design was still being refined. The Air Force accepted the proposal only because Convair was to foot the bill, which would not be cheap. The cost of constructing and testing the two B-60 prototypes would eventually come to $15 million.

In late summer 1951, Convair revealed its second design. It was basically similar to the YB-36G model, with the exception of a lengthened airframe by 2.5 ft. to 172.1 ft. The USAF was indifferent, and the overpowering presence of the YB-60's challenger, the YB-52, was all-pervasive. Convair was feeling the pressure.

The Convair design team did well to come up with another alternate YB-60 design on July 11, 1952. However, by that time, it was too late. The design Convair proposed this time placed increased emphasis on defensive armament, while overall performance did not improve and, in a way, deteriorated. That caused further worsening of the gulf, regarding performance, between the Convair design and the Boeing YB-52. The July 1952 design had the same height, wing span, and wing area as the previous two designs, but the YB-60's length was further increased by another 4.1 ft. That brought its length to 175.2 ft. The problem was that the eight YJ57-P-3 turbojet engines still powered the aircraft with the same ratings as previously.

Also, the increased heavy weaponry on the YB-60, for example, ten M24A1 20 mm cannon, and a load of 3,600 rounds, added weight to the large airframe. Furthermore, there was an increase in crew numbers. Whereas the previous two designs had only five members, the July 1952 variant needed almost double the amount of crew, nine in all.

These new additions to the bomber increased the weight of the new design, affecting performance heavily. For instance, a ground run during an unassisted take-off increased from 6,710 ft. to 8,000 ft., with an altitude of 50 ft. was attained at a distance of 9,700 ft. That was 1,569 ft., more than the 1951 summer design. Maximum speed also dropped from 515 mph to 507 mph at an altitude of 39,250 ft.

Boeing's YB-52 was actually the USAF's choice from the beginning. But, unfortunately, the YB-60 was like the ugly duckling in that love triangle.

Things never got off to a good start for the Convair colossus. When the first YB-60 conducted its maiden flight from Carswell AFB, Fort Worth, Texas, on April 18, 1952, the YB-52 had already made a successful maiden flight three days previously. From then on, it was all downhill for the YB-52 and a very grueling climb up the hill for the YB-60.

THE ROAD TO FAILURE - YB-60

APRIL 1952

Of the first of two Convair prototype YB-60s ordered, the first was practically finished in October 1951, except that it didn't have power plants. In fact, it never got its engines and never flew. The second was the B-36 advanced pure-jet swept-wing development, which included eight Pratt & Whitney J-57S turbojets. As with the standard B-36, it was built with 9,000 pounds of magnesium. The inner wings were similar to regular B-36 wings, while the outside panels had a 350 sweep. In particular, the fuselage between the cockpit and a point near the tail remained unchanged. The nose was modified, while a tail parachute was installed at the back end. Around 72% of the structure was identical to the previous B-36. That was the design proposed to the USAF in July 1952, which ultimately failed.

JULY 1952

After the completion of flight trials by the YB-60 and the YB-52, the USAF made its choice. In late July 1952, the Air Force decided to concentrate on the Boeing B-52 and eliminate the Convair YB-60, which meant the Convair bomber would not go into production, even though such an idea had been under consideration. The limited funds the USAF had at its disposal for heavy bomber equipment necessitated concentration on one long-range bomber, and the Air Force could not go past the B-52. The decision by the USAF cleared the way for concentration on accelerated production of the Boeing B-52.

JANUARY 1953

In one last hurrah, the YB-60 landed at Edwards AFB, California, on January 7, after completing an impressive 1,050-mile non-stop flight from Carswell AFB, Texas. The swept-wing version of the B-36 flew to California for engine calibration tests.

AUGUST 1953

A little over one year after the YB-60 lost its battle against the YB-52, it faced another hurdle. There were discussions between the USAF and Convair to use the Convair YB-60 airframes for accelerated testing of the P&W J57 turbojets. However, that plan was dropped by the Air Force. Instead, the USAF directed Convair to mothball the two prototype YB-60 jet bombers. Hence, it was the end of the road for the prototype that never flew because the USAF would not supply engines for it. The other one did take to the skies, but it only logged about 40 hours. As a result, no further development was made with the two prototype Convair YB-60s.

Since 1952, the two prototype YB-60s had been used as test beds for various engines, electronic gear, and other devices, including a unique Convair 'diamond-form' undercarriage. Unfortunately, they didn't last long, even in those endeavors. They were both unceremoniously scrapped at Carswell AFB in July 1954.

A top side view of YB-60, 49-2676 (USAF Museum)

The YB-60 flying into Rogers Dry Lake, California. (USAF, AFFTC/HO, Edwards AFB)

Even though the YB-60 lost the competition with the YB-52, it was a sleeker, and better looking plane than the BUFF. (T. Gianna, USAF)

The YB-60 taking a break on the tarmac with a trio of B-36s at the back. (T. Gianna, USAF)

The Convair YB-60 on its third flight over a rural area. (T. Gianna, USAF)

YB-60 Specifications (SAC model)

TYPE	POWERPLANT	DIMENSIONS WINGSPAN / LENGTH / HEIGHT	WEIGHTS Empty / Combat / Gross / Max. t.o.w.	SPEEDS Cruise / Average Speed for Range / Max speed	SERVICE CEILING / RANGE
YB-60	8 x Pratt & Whitney XJ57-P-3 turbojet engines with 8,700 lbf (39 kN) thrust	206 ft / 171 ft / 50 ft	153,000 lb. / 260,250 lb. (Basic Mission) / 300,000 lb. / 346,000 lb. (Normal), 410,000 lb. (Overload) (Limited by performance)	407 mph Basic Mission / 188 mph Ferry Mission / 508 mph (at optimal altitude of 39,250 ft.)	52,000 ft. / 10,000 miles max ferry range
YB-60 WEAPONS	Two 20mm cannon in the tail, and up to 72,000 lbs. of bombs				

B-36 GALLERY

The single prototype XB-36, 42-13570, getting ready for takeoff. Note the gigantic 110 inch wheels (USAF)

A B-36 on the ground at an undisclosed airfield. (National Archives)

Four B-36 Peacemakers approaching Chicago over Lake Michigan, on July 3, 1949. (USAF - AFMC History & Heritage Program)

The B-36, a popular attraction at any airshow, drew enormous crowds wherever it went, including this Open House in El Paso, Texas, on December 5, 1948. (USAF - AFMC History & Heritage Program)

Night View of a B-36 in a hangar, in1955. (National Archives)

An Oregon Air National Guard F-94B, 51-5450, escorts Convair B-36D, 44-92033. (USAF - AFMC History & Heritage Program)

B-36 'Four Deuces' flying over the Atlantic on October 28, 1956 with number one prop feathered. (USAF - AFMC History & Heritage Program)

Consolidated Vultee B-36 Formation over Fort Sheridan, Illinois during the Chicago Air Fair in1949. (National Archives)

A Convair B-36 used by the Special Weapons Command in March, 1953. Note the Republic F-84 on the right. (National Archives)

A lone B-36 flies above some town, in some state of America, letting the populace know it's got their backs. (National Archives)

This is a poignant photo showing a B-52 scooting away from a slow-moving B-36. It is vivid enforcement of the differences between the two bombers. One was already an archaic World War II concept in the mid-1950s, while the other was a trendsetting jet bomber conceived in the golden years of the rocking 1950s. Speaking of Rock, the B-52 is still rocking hard in the 2020s, and could be the first bomber in the world to stay in service for 100 years. After undergoing major upgrades between 2013 and 2015, the B-52 is on course to attain that record in 2055. (National Archives)

SERIAL NUMBERS - B-36, XC-99, YB-60	
XB-36	42-13570
B-36A	44-92004/92025
B-36D	49-2647/2668
B-36F	49-2669/2675; 2677/2683; 2685; 50-1064/1082
RB-36F	49-2703/2721; 50-1098/1102
B-36H	50-1083/1097; 51-5699/5742; 52-1343/1366
RB-36H	50-1103/1230; 51-5743/5756; 13717/13741; 52-1367/1392
YB-36	42-13571
B-36B	44-92026/92098
RB-36D	49-2686/2702
YB-60	49-2676 & 49-2684
XC-99	43-52436

CONVAIR B-36, XC-99, YB-60 PRODUCTION LIST	
MODEL	UNITS PRODUCED
XB-36	1
YB-36	1
B-36A	22
B-36B	62
B-36D	26
RB-36D	24
B-36F	34
RB-36F	24
B-36H	83
RB-36H	73
B-36J	33
XC-99	1
YB-60	2
TOTAL	385

Other Image Credits

Page **36, Hunini,** CC BY-SA 4.0 <https://creativecommons.org/licenses/by-sa/4.0>, via Wikimedia Commons (https://commons.wikimedia.org/wiki/File:General_Electric_J47-GE 27_turbojet_engine_left_front_view_at_Hamamatsu_Air_Base_Publication_Center_November_24,_2014.jpg)

Page **61 Bill Larkins,** "B-36atSFO" Flickr photo by Bill Larkins https://flickr.com/photos/34076827@N00/1003222173 shared under a Creative Commons (BY-SA) license 2.0

Page **69 Mike LaChance,** "CONVAIR B-36J Peacemaker" Flickr photo by mikelachance816 https://flickr.com/photos/mjl816/46676712024 shared under a Creative Commons (BY) license 2.0

Made in the USA
Las Vegas, NV
03 November 2026

57424065R00057